Complex, Hypercomplex and Fuzzy-Valued Neural Networks

Complex, Hypercomplex, and Fuzzy-Valued Neural Networks are extensions of classical neural networks to higher dimensions. In recent decades, this theory has emerged as a forefront in neural networks theory. There are several approaches to extend classical neural network models: quaternionic analysis, which merely uses quaternions; Clifford analysis, which relies on Clifford algebras; and finally generalizations of complex variables to higher dimensions. This book reflects a selection of papers related to complex, hypercomplex analysis, and fuzzy approaches applied to neural networks theory. The topics covered represent new perspectives and current trends in neural networks and their applications to mathematical physics, image analysis and processing, mechanics, and beyond.

Complex, Hypercomplex and Fuzzy-Valued Neural Networks

New Perspectives and Applications

Agnieszka Niemczynowicz,
Irina Perfilieva, Lluís M. García-Raffi,
and Radosław Kycia

Routledge
Taylor & Francis Group

NEW YORK AND LONDON

Designed cover image: Shutterstock

First edition published 2026
by Routledge
605 Third Avenue, New York, NY 10158

and by Routledge
4 Park Square, Milton Park, Abingdon, Oxon, OX14 4RN

Routledge is an imprint of Taylor & Francis Group, LLC

ISBN: 978-1-032-84714-6 (hbk)
ISBN: 978-1-032-84851-8 (pbk)
ISBN: 978-1-003-51530-2 (ebk)

DOI: 10.1201/9781003515302

Typeset in Nimbus Roman font
by KnowledgeWorks Global Ltd.

Publisher's note: This book has been prepared from camera-ready copy provided by the authors.

Contents

CHAPTER   3 ▪ Applications of Genetic Algorithms in Neural Networks   **40**

WOJCIECH KSIĄŻEK

SECTION III Theoretical Foundation of Computation with Neural Networks

Preface

This book was written by a specialist carefully selected to cover a wide range of topics related to real, hypercomplex, and fuzzy neural networks. The selection process aimed to ensure both deep theoretical knowledge and practical expertise so that the material presented here would be not only comprehensive but also accessible to a broad audience. We believe that a book of this kind is not yet available on the market, and this gap deserves to be filled—especially at a time when neural network architectures are rapidly evolving and expanding into new mathematical domains.

Our goal was to create a resource that bridges the gap between classical neural network theory and emerging approaches that leverage more abstract number systems and fuzzy logic. By doing so, we hope to inspire further research and innovation in this exciting and fast-moving field. We also aimed to make the content suitable for both researchers and advanced students who wish to explore the frontiers of deep learning.

We hope that our passion for deep learning research will be beneficial to you, dear Reader, and that this book will serve as a valuable reference and source of inspiration throughout your journey in this area.

Introduction

This book brings together a collection of chapters dedicated to a wide spectrum of topics within the field of neural networks, with a particular focus on real-valued, hypercomplex (including complex and quaternionic), and fuzzy neural networks. The structure of the book has been carefully designed to reflect this diversity, dividing the content into three comprehensive sections: Real-Valued Neural Networks, Complex- and Quaternionic-Valued Neural Networks and Their Applications, and Theoretical Foundations of Computation with Neural Networks, which also incorporates fuzzy modeling approaches. Each section presents contributions written by recognized experts in the respective subfields, ensuring both depth and relevance of the material.

The first part is devoted to real-valued neural networks, the most well-established and widely studied category in the field. It starts with fundamental principles and architectures associated with classical neural networks, providing a strong theoretical grounding. However, it goes beyond the basics to explore advanced and contemporary directions in the area, such as the development and functioning of Large Language Models (LLMs), which represent the cutting edge of natural language processing. This part also includes work on the application of evolutionary computation techniques, particularly genetic algorithms, to neural network training and optimization—demonstrating the power of hybrid approaches that combine different computational paradigms.

The second part shifts the focus toward more abstract numerical representations by exploring complex-valued and quaternionic-valued neural networks. These approaches are motivated by the need to model multidimensional data more naturally and efficiently. The section begins with accessible yet rigorous introductions to the underlying algebraic structures of complex and quaternion numbers, making it suitable even for readers less familiar with these mathematical systems. It then discusses how these algebras can enhance the representation power of neural networks, especially in domains where phase, rotation, or multidimensional encoding plays a central role. Practical applications are also included—such as encoding genetic information using quaternionic neural networks and advanced image analysis tasks like melanoma detection—highlighting the growing relevance of these methods in real-world problems.

The third and final part addresses the theoretical foundations of neural network computation. This section delves into the mathematical underpinnings that support the expressive power and approximation capabilities of neural networks, including a discussion of Cybenko's universal approximation theorem and its generalizations to broader contexts. In addition to these theoretical insights, the section presents the integration of fuzzy logic into neural computation. Fuzzy modeling techniques are particularly valuable when dealing with imprecise, uncertain, or time-dependent data, and the presented work illustrates their application to tasks such as time series prediction and analysis.

Taken together, the chapters in this book provide a rich and multifaceted exploration of the current landscape of neural network research. By combining foundational theory with practical applications and extending beyond conventional real-valued models, this volume offers a unique and timely perspective. We believe that the content will be of interest not only to researchers and practitioners working in artificial intelligence and machine learning but also to advanced students seeking to expand their understanding of emerging directions in deep learning.

We hope that the depth, diversity, and interdisciplinary spirit of this book will inspire new ideas and further exploration in the ever-evolving world of neural computation.

ACKNOWLEDGEMENT

The work on the book has been supported by the Polish National Agency for Academic Exchange Strategic Partnership Programme under Grant No. BPI/PST/2021/1/00031.

I

Real-Valued Neural Networks

## CHAPTER 1

Theoretical Foundation of Real-Valued Neural Networks

Radosław Kycia, Agnieszka Niemczynowicz

Faculty of Computer Science and Telecommunications, Cracow University of Technology, Warszawska 24, 31-155 Kraków, Poland.

1.1 INTRODUCTION

In the past two decades, neural networks have gained significant interest as a convenient tool useful for solving a wide variety of practical tasks. They have been successfully applied to an extremely broad range of problems in diverse fields such as finance [30], medicine [171], engineering, geology, and physics [23]. Neural networks are used wherever issues related to data processing and analysis arise, including prediction, classification, or control tasks. Neural networks are a highly sophisticated modeling technique, capable of representing exceptionally complex functions. Their non-linear nature significantly enriches the scope of their applications.

For many years, linear modeling was the commonly used mathematical technique for describing various objects and

DOI: 10.1201/9781003515302-1

3

processes. Optimization strategies for building such models are well-established. However, in many cases, there is no basis for applying linear approximations to a given problem, and linear models often fail, leading to premature conclusions about the 'impossibility' of mathematically describing a particular system. In such instances, using models created with neural networks can be the quickest and most convenient solution. Neural networks also enable control over the complex problem of high dimensionality, which, with other approaches, significantly complicates attempts to model non linear functions with a large number of variables. In practice, neural networks automatically create the models needed by the user, as they learn from the examples provided to them.

One needs some (mainly empirical) knowledge regarding the selection and preparation of training data, has to make the right choice of neural network architecture, and has to be able to interpret the results. However, the level of theoretical knowledge required to effectively build a model is significantly lower when using neural networks compared to traditional statistical methods or mathematical modeling.

A very interesting feature of neural networks stems from the fact that they, to some extent, mimic the functioning of the human brain. These networks are based on a very simple model, representing only the most basic essence of how the biological nervous system operates, but it is an attempt to penetrate the core of its functioning. Some believe that the development of neurobiological modeling could lead to the creation of truly intelligent computers, endowed with initiative and capable of making independent decisions.

What makes this rather 'exotic' computational tool so popular? Neural networks can be applied with a high probability of success in cases where there are problems in creating mathematical models. Through the learning process, they allow the automatic mapping of various complex relationships between input and output signals. It is certainly justified to speak of neural networks as a very interesting and modern method for solving problems, with still untapped potential.

The technique of neural networks is no longer a special novelty today. It can be assumed that the field itself truly emerged with the publication of the historic paper by [97], where, for the first time, an attempt was made to mathematically describe a nerve cell and link this description to the problem of data processing.

Neural networks were developed as a result of research conducted in the field of artificial intelligence, with particular significance given to studies that focused on building models of the fundamental structures found in the brain.

Research conducted in the field of so-called symbolic artificial intelligence, during the years 1960-1980, led to the development of so-called expert systems. These systems are based on a general model of the formalized reasoning process. While very useful in certain areas, these systems were unable to mimic certain elementary structures in the human brain and thus could not explain key aspects of human intelligence. This led to the belief that in order to build a fully intelligent system, it has to be modeled after the structure of real intelligent systems, i.e., the structure of the brain.

The human brain (based on anatomical and histopathological research) consists primarily of a large number of elementary nerve cells, called neurons. It is estimated that there are about 10 billion neurons, most of them interconnected in the form of a complex network. It has been established that, on average, each neuron has several thousand connections, although the number of connections for individual cells may vary.

Each neuron is a specialized biological cell capable of transmitting and processing complex electrochemical signals. A neuron typically has a branched structure of input signals (dendrites), a body (perikaryon) that integrates signals from all inputs, and an output carrier of information (axon), which exits the cell as a single fiber and then multiplies the signal it processes, sending it to various recipient neurons through a branched output structure (telodendron).

The axon of one cell connects with the dendrites of other cells through biochemical junctions that modify signals and serve as memory carriers. These are the so-called synapses, which, in their biological form, are highly complex. However, in artificial neural networks, they are simplified to operators that multiply input signals by coefficients determined during the learning process. A neuron activated by synapses enters an active state, which manifests as the transmission of an electrochemical signal through its output axon. This signal has a characteristic shape, amplitude, and duration, and it reaches other neurons through subsequent synapses.

A neuron only becomes excited if the combined signal received by the cell body through its dendrites exceeds a certain threshold level. The strength of the signal received by the neuron largely depends on the effectiveness (or weight) of the synapse through which the impulse reaches it. Each synapse has a gap that is filled with a special substance called a neurotransmitter or neuromediator when a signal between neurons is transmitted. The mechanism by which the neurotransmitter (*the synapse weight*) functions is of great significance in the biology of the nervous system, as the changes in neurotransmitter chemistry can artificially influence human behavior. Therefore, neurotransmitter concentrations influence the transmission (and either amplify or weaken) of a signal across the gap in each synapse between the signaling neuron and the receiving neuron.

One of the most well-known researchers of neurological systems, Donald Hebb, proposed that the learning process primarily involves changes in the 'strength' of synaptic connections. In the classic experiment by Pavlov on conditioned reflexes, where a bell rings before a dog is fed, the dog quickly learns to associate the sound of the bell with food. This occurs as specific synaptic connections are strengthened as a result of the learning process.

It is now believed that by using a large number of these simple learning mechanisms, along with numerous but very simple information-processing elements, such as neurons, the brain is capable of performing the incredibly complex tasks it executes daily. Of course, in the actual biological brain, many more complex information-processing mechanisms are at work, involving additional elements.

The crucial role in the computing by the neural networks is played by the topology of connections between various neurons. Modifying these connections alters both the hardware and the software. This approach is called connectionism. The human brain has some, so-called, neuroplasticity, which is the ability to change connections, resulting in reprogramming or adjusting the brain to new tasks.

In order to replicate only the basic structure of biological nervous systems, the creators of artificial neural networks decided to define an artificial neuron as follows:

- A certain number of input signals (values) reach the neuron.

- Each value is introduced to the neuron through a connection with a certain strength (weight). These weights correspond to the efficiency of the synapse in a biological neuron.

- Each neuron also has a single threshold value, determining how strong the stimulation must be for the neuron to be activated.

- In the neuron, a weighted sum of inputs is calculated (i.e., the sum of input signal values multiplied by their respective weight coefficients), and then the threshold value is subtracted from it. The resulting value determines the neuron's activation. This is, of course, a highly simplified approximation of actual biological phenomena.

- The signal representing the neuron's overall activation is then transformed by a set activation function (the neuron's transfer function). The value calculated by the activation function becomes the final output (output signal) of the neuron.

The behavior of a neuron (and the entire neural network) is highly dependent on the type of activation function used. It is very interesting, and even intriguing, that artificial neural networks can achieve such significant practical results using an extremely simplified model of a neuron. The complexity of this model is not much greater than the scheme where the neuron simply calculates the weighted sum of its inputs and activates when the total input signal exceeds a certain threshold level. In artificial neural networks, activation functions that provide continuously varying signals are commonly used. The most frequently used activation functions are in the form of the so-called sigmoid function.

In the artificial neural network model, input signals often include both positive and negative values (excitation and inhibition). This is meant to model the actual brain's excitatory and inhibitory pathways (which, in reality, are separate and carried out by specialized inhibitory neurons). For a neural network to be useful, it must have inputs (used to transmit externally observed variable values) and outputs (which represent the results of computations).

Inputs and outputs correspond to specific nerves in the brain: sensory nerves for inputs and motor nerves for outputs. There can also be neurons that perform internal functions within the network, which mediate the analysis of information delivered by sensory nerves and participate in processing sensory signals into decisions that activate specific executive elements.

Since an external observer does not have access to the inputs or outputs of these intermediary neurons, they are typically referred to as hidden neurons. Hidden neurons (or hidden layers) are those elements of the network to which signals cannot be directly transmitted or received from either the input or the output side.

Input, hidden, and output neurons must remain interconnected, which involves the problem of choosing the network architecture. A key issue in selecting the network architecture is the presence or absence of feedback loops in the structure.

Simple networks have a unidirectional structure. Signals flow only in one direction — from inputs, through successive hidden neurons, ultimately reaching the output neurons. Such a structure is characterized by stable behavior, which is an advantage. The network can also have built-in feedback loops (i.e., connections that return from later neurons to earlier ones), allowing it to perform more complex computations, particularly those with a recursive nature. Research shows that a network with fewer neurons that includes feedback loops can perform computations as complex as those of a network without feedback loops but with a significantly larger number of neurons.

However, this does not come without any 'costs' — due to the circulation of signals in networks with feedback (from input to output and back to input through feedback), the network can exhibit unstable behavior and have very complex dynamics, potentially leading to the most complicated forms of behavior, such as deterministic chaos.

Neural networks with a relatively large number of feedback connections, specifically networks where all connections are feedback connections, exhibit considerable practical usefulness. Such networks are known as Hopfield networks [66]. The operation of a neural network results from the actions of individual neurons and the interactions between them. From a mathematical perspective, a single neuron typically performs the operation of the dot product between the vector of input signals and the weight vector. As a result, the neuron's response depends on the geometric relationships between the signal vectors and the weight vectors. The correct geometry of the weight vectors, which ensures proper functioning, is achieved through a learning process. This process can be interpreted as an automatic method for finding a set of weight coefficients across all neurons in the network that guarantees the smallest total error made by the network.

By using an appropriate learning algorithm (the most well-known being the backpropagation algorithm), the network can systematically reduce the error made during the learning process, resulting in gradual improvement in its performance over time.

Although neural networks are a completely independent field of knowledge, in practical applications, they usually serve as the controlling or decision-making part, providing executive signals to other components of the 'device' that are not directly related to neural networks. The functions performed by neural networks can be categorized into several basic groups: approximation and interpolation, pattern recognition and classification, compression, prediction and control, and association. In each of these applications, a neural network acts as a universal function approximator for multiple variables [41], implementing a nonlinear function of the form $y = f(x)$, where x is the input vector and y is the vector function of multiple variables. A large number of modeling, identification, and signal processing tasks can be reduced to approximation problems. In classification and pattern recognition, the network learns the basic features of these patterns, such as the geometric mapping of the pattern's pixel arrangement, the distribution of principal components of the pattern, Fourier transform components, or other properties. The learning process emphasizes the differences among various patterns, which form the basis for assigning them to the appropriate class.

In the field of prediction, the network's task is to determine the future responses of a system based on a sequence of past values. Given information about the values of variable x at moments preceding the prediction $x(k-1), x(k-2) \ldots x(k-N)$, the network decides what the estimated value $x(k)$ of the sequence will be at the current time k. The adaptation of network weights utilizes the current prediction error and the value of that error at previous moments.

In dynamic process identification and control problems, a neural network typically performs several functions. It acts as a nonlinear model of the process, enabling the development of an appropriate control signal. It also functions as a tracking and adaptive system, adjusting to environmental conditions. An important role, especially in robot control, is played by the classifier function used in making decisions about the subsequent course of the process. In associative tasks, a neural network functions as associative memory. There are two types of associative memory: associative memory, where associations involve only individual components of the input vector, and hetero-associative memory, where the network's task is to associate two vectors with each other. If a distorted vector (e.g., with elements corrupted by noise or missing certain data elements) is presented as input to the network, the neural network can reconstruct the original vector, free from noise, and generate the complete form of the vector associated with it.

The most significant feature of neural networks, which contributes to their considerable advantages and wide range of applications, is the parallel processing of information by all neurons. This parallelism, combined with a large scale of neural connections, significantly accelerates the information processing process. In many cases, real-time signal processing is possible. The extensive number of inter-neuronal connections makes the network robust against errors occurring in some connections. Damaged weights are taken over by others, and as a result, significant disturbances are not noticeable in the network's performance. These properties are utilized, among other things, in the search for the optimal neural network architecture by pruning certain weights.

It is also visible in LLMs (Large Language Models) that the great power of neural networks in generating responses in natural language also has its drawback — LLMs hallucinate, i.e., provide untrue answers. This situation is an obvious result of asking about information that is not in the training data. That behavior also suggests that ANN systems must be somehow supervised or should be able to explain the result (XAI, *Explainable Artificial Intelligence*). These are current topics of research.

1.2 SINGLE NEURON MODELS

A neuron is the fundamental information-processing unit crucial for the functioning of neural networks. The neuron model was first formally defined by W.S. McCulloch and W.H. Pitts, who were pioneers in this field in their work [97]. *The formal neuron model* consists of weights, input activations, a total activation function, and a transfer function, often referred to as a threshold function. It also includes an external bias that affects the net input to the activation function, either increasing or decreasing it depending on whether the bias is positive or negative. The transfer function determines how the neuron processes the incoming signal before passing it to the activation function. In typical neuron models the transfer function is a linear operation, which calculates the sum of the weighted inputs along with the bias. The bias is an extra parameter added to the transfer function, allowing the activation function to shift. This adds flexibility to the neuron, enabling it to activate or deactivate even when the input values are near zero, especially when the weights of the input signals are small. From a mathematical point of view, the kth neuron can be defined as [97, 61]

$$u_k = \sum_{j=1}^{n} w_{kj} x_j, \tag{1.1}$$

$$y_k = \phi(u_k + b_k), \tag{1.2}$$

where x_j, for $j = 1, \ldots, n$ are the inputs signals, w_{kj} are weights of neuron k.

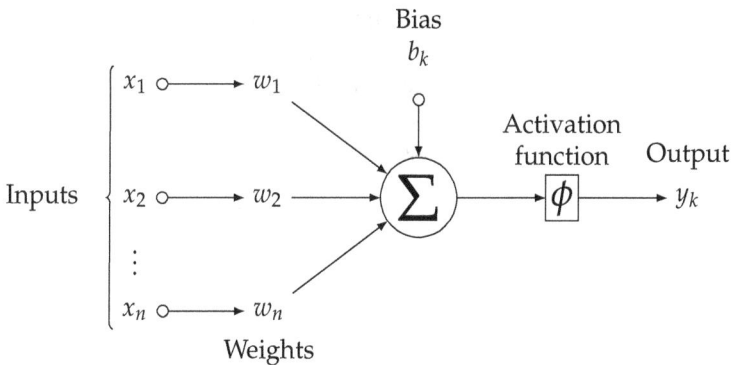

Figure 1.1 Structure of neuron.

Here u_k is a linear combination of inputs and their weights, b_i is a bias, ϕ is an activation function, and y_k is the output signal of neuron k.

1.3 ACTIVATION FUNCTIONS

Activation functions play a significant role in determining the computational complexity of neural networks, as they introduce non-linearity, enabling the model to capture complex patterns in data [89]. While we will present some popular examples, it is important to remember that there is a vast range of activation functions available, and their selection typically depends on the specific requirements of the research problem being addressed by the neural network.

One of the simplest activation functions is the *linear function*, defined as $f(x) = x$. While easy to implement, it lacks non-linearity, restricting the network to linear behavior.

The *sigmoid function*, given by $f(x) = 1/(1 + e^{-x})$, outputs values in the range $(0, 1)$. It is commonly used for binary classification but suffers from the vanishing gradient problem for large or small input values, which can slow down learning. Another widely used activation is the *hyperbolic tangent* (tanh), defined as $f(x) = \tanh(x) = (e^x - e^{-x})/(e^x + e^{-x})$, with a range of $(-1, 1)$. Its symmetry around zero aids optimization, but it can also suffer from the vanishing gradient problem.

The *Rectified Linear Unit (ReLU)*, expressed as $f(x) = \max(0, x)$, is computationally efficient and addresses the vanishing gradient issue. However, it may cause the *dying neurons problem*, where neurons output zero for non-positive inputs. An improvement over ReLU is the *Leaky ReLU*, defined as $f(x) = x$ for $x > 0$ and $f(x) = \alpha x$ for $x \leq 0$, where α is a small positive constant. This function mitigates the dying neurons problem but requires careful tuning of α.

The *Exponential Linear Unit (ELU)* is another alternative, defined as:

$$f(x) = \begin{cases} x & \text{for } x > 0, \\ \alpha(e^x - 1) & \text{for } x \leq 0, \end{cases}$$

where $\alpha > 0$. ELU reduces bias shift and improves learning but involves a higher computational cost than ReLU.

For multiclass classification, the *softmax function*, given by $f(x_i) = e^{x_i} / \sum_j e^{x_j}$, is often used. It outputs values in the range $(0, 1)$, with their sum equal to 1. However, it is sensitive to large input values, which may lead to numerical instability.

Recent advancements include *Swish*, defined as $f(x) = x \cdot \sigma(x)$, where $\sigma(x) = (1 + e^{-x})^{-1}$. Swish provides smooth activation and performs well in deep networks. Another is the *Gaussian Error Linear Unit* (GELU), which combines the properties of ReLU and sigmoid, offering smooth transitions and improved performance.

Activation functions play a vital role in determining the efficiency and accuracy of neural networks. The choice of function depends on the problem, network architecture, and data.

1.4 HOW PERCEPTRON WORKS?

In this section we describe in detail the way of how a perceptron (neuron unit) works. This is extremely important to understand how a connected system of neurons works, at least in principle.

The general idea of perceptron work is related to the separation of *decision regions* by an $(n-1)$-dimensional hyperplane in an n-dimensional data space. The decision regions are domains in the data space that are labeled by a specific data etiquette. The data that falls into the specific decision region are classified according to the given label.

Figure 1.2 represents data that can be *linearly separable*; it means that the data of different labels can be separated by a hyperplane (i.e., a line in the plane). Therefore, the idea of separa-

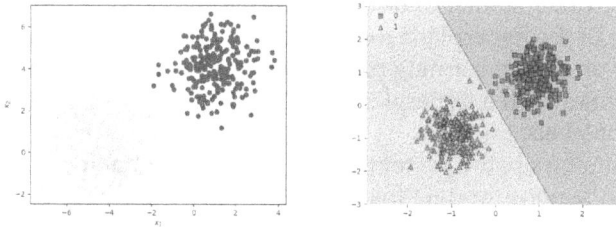

Figure 1.2 Left: data points with two labels marked by different colors. Data are visually separated. Right: decision regions created by split data plane by a line.

tion of decision regions by the hyperplane works only for specific 'topology' of data — linearly separable data. The perceptron is *the supervised classification algorithm* that uses a hyperplane to separate the decision region. It can be seen by the following computation:
(1) Define the weight vector with the bias by $\vec{w} = [b \ w_1 \ \ldots \ w_n]^T$ and the data vector augmented by 1 in the corresponding bias

position in the form $\vec{x} = \begin{bmatrix} 1 & x_1 & \dots & x_n \end{bmatrix}^T$, which gives

$$\phi(\vec{x}, \vec{w}) := \vec{x} \cdot \vec{w} = \sum_{i=1}^{n} x_i w_i + b.$$

(2) Fixing $\vec{w}$ and varying $\vec{x}$ over[1] $\mathbb{R}^n$, by the equation

$$\phi(\vec{x}, \vec{w}) = 0, \tag{1.3}$$

we define an $(n-1)$-dimensional hyperplane in an n-dimensional linear space where data point. Let us apply the simplest and prototypical activation function, i.e., theta Heaviside's distribution[2]

$$\theta(x) = \begin{cases} 1 & \text{for} \quad x > 0, \\ 0 & \text{for} \quad x \le 0. \end{cases} \tag{1.4}$$

Taking activation function $\sigma(x)$ as $\sigma(x) = \theta(x)$ and applying it to the hyperplane equation, we have that $\sigma \circ \phi$ is 1 for $\vec{x}$ lying on one side of the hyperplane (1.3) and is 0 for those $\vec{x}$ that lie on a hyperplane or on the opposite side of the hyperplane from those $\vec{x}$ for which $\sigma \circ \phi$ attains value 1. This means that $\sigma \circ \phi$ discriminates the side of the hyperplane fixed by $\vec{w}$.

The process of learning of a perceptron is realized by fixing $\vec{w}$ that defines such a hyperplane that best separates data with two different labels.

Summing up, the perceptron is a good algorithm for devising decision regions for data that can be linearly separated. For data that are not linearly separable, see e.g., Fig. 1.3, the perceptron works with non perfect accuracy — some points fall in the wrong decision region.

One prototypical and simple example that illustrates the inability of a perceptron to learn not linearly separable data is the XOR (eXclusive OR) gate, which is defined by the function

---

[1]$\vec{x} \in \mathbb{R}^{n+1}$; however, the first component is fixed to 1, so effectively the n-components: $\{x_i\}_{i=1}^{n}$ can vary over $\mathbb{R}^n$.

[2]Heaviside's theta function/distribution is a discontinuous function at $x = 0$. Its value at $x = 0$ can be arbitrary choose, e.g., $\theta(0) = \frac{1}{2}$ or $\theta(0) = 1$. This function is also important in electrotechnics applications and is used to define the basic example of so-called Lauren Schwartz distributions, i.e., linear functional over the space of test functions.

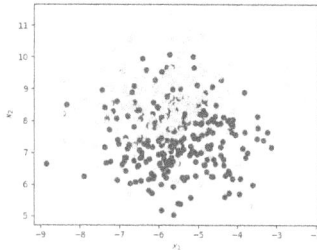

Figure 1.3 Data that cannot be linearly separated.

(False $\to$ 0, True $\to$ 1):

x	y	$XOR(x,y)$
0	0	0
1	0	1
0	1	1
1	1	0

(1.5)

If one treats the value of the XOR function as a label 0 or 1, then the data points (x,y) are not linearly separable, and from the above mathematics, the perceptron cannot describe proper decision regions, which roughly can be summarized as: 'perceptron cannot learn XOR gate.' The large investment in this research direction and disappointments from this technology at the initial stage of development led to the break in funding and, as a result, in research, in this direction called the first AI winter in the 1970s.

The activation function σ can also be chosen as a continuous and an even differentiable or a smooth function. However, one of the crucial properties of this function is to discriminate between the different sides of the hyperplane.

1.5 HOW PERCEPTRON LEARNS?

Perceptron is a supervised learning algorithm. Therefore, we have data that can be split into two parts:

- features/attributes/measures: $\vec{x}_i \in \mathbb{R}^{n_x}$ for $i = 1 \ldots N$;

- class labels/targets: $\vec{y}_i \in \mathbb{R}^{n_y}$ for $i = 1 \ldots N$;

The joint vector $(\vec{x}, \vec{y})$ is called a sample or an observation.

To formalize the learning process, we must choose a distance/metric[3] that measures the difference between the output of the perceptron $\sigma \circ \phi(\vec{x}_i, \vec{w}) = \vec{y}_i$ for the input $\vec{x}_i$ and the true label $\vec{y}_i$ for all $i \in 1 \ldots N$. We define it as:

$$d : \mathbb{R}^{n_y} \times \mathbb{R}^{n_y} \to \mathbb{R}_{\geq 0} := \{x \in \mathbb{R} | x \geq 0\}, \qquad (1.6)$$

which typically should fulfill the typical metric definition:

- $d(\vec{y}, \vec{y}) = 0$,
- $d(\vec{y}_1, \vec{y}_2) > 0$ if $\vec{y}_1 \neq \vec{y}_2$,
- $d(\vec{y}_1, \vec{y}_2) = d(\vec{y}_2, \vec{y}_1)$ (symmetry),
- $d(\vec{y}_1, \vec{y}_2) \leq d(y_1, \vec{y}_3) + d(\vec{y}_3, \vec{y}_2)$ (triangle inequality).

The distance can be defined using a norm

$$||\_|| : \mathbb{R}^{n_y} \to \mathbb{R}_{\geq 0}, \qquad (1.7)$$

with the following properties

- $||\vec{y}|| = 0$ if and only if $\vec{y} = 0$,
- $||\lambda \vec{y}|| = |\lambda| \cdot ||\vec{y}||$ for $\lambda \in \mathbb{R}$,
- $||\vec{y}_1 + \vec{y}_2|| \leq ||\vec{y}_1|| + ||\vec{y}_2||$ (triangle inequality).

The example norm are:

- p-norm: $||\vec{x}||_p = \sqrt[p]{\sum_{i=1}^{n} |x_i|^p}$,
- Euclidean norm: $||\_||_2$,
- Manhattan/taxicab norm: $||\_||_1$,
- maximum norm: $||\vec{x}||_\infty := max_{i \in \{1,\ldots,n\}} (|x_i|)$,

for a vector $\vec{x} = [x_1, \ldots, x_n] \in \mathbb{R}^n$.

The metric/distance is connected with the norm by the following relation

$$d(\vec{y}_1, \vec{y}_2) := ||\vec{y}_1 - \vec{y}_2||. \qquad (1.8)$$

---

[3]Usually these two terms are used interchangeably.

The overall error that is made by perceptron can be characterized by

$$err(\{\vec{x}_i, \vec{y}_i\}_{i=1}^N, \vec{w}) := \sum_{i=1}^N d(\vec{\bar{y}}_i, \vec{y}_i) = \sum_{i=1}^N d(\sigma \circ \phi(\vec{x}_i, \vec{w}), \vec{y}_i). \quad (1.9)$$

Instead of d we can use $f \circ d$ for any strict monotone function, e.g., $f(x) = x^2$ that makes $err(\ldots)$ function differentiable[4].

We can now formulate general learning process of perceptron as an optimization problem:

Definition 1. The perceptron $\sigma \circ \phi(\_, \vec{w})$ is trained by selecting the weights $\vec{w}$ in the optimization process

$$min_{\vec{w} \in \mathbb{R}^{n+1}} err(\{\vec{x}_i, \vec{y}_i\}_{i=1}^N, \vec{w}). \quad (1.10)$$

One can note that in the optimization process we try to find $\vec{w} = \vec{w}(\{\vec{x}_i, \vec{y}_i\}_{i=1}^N)$, and different training data leads, in general, to different values of $\vec{w}$.

The optimization process can be realized in many ways [81]. One of the simplest ways, which is in the direction of the back-propagation algorithm for multilayer perceptrons, is the gradient descent method. To find a minimum of $f(\vec{w})$, we start from the initial value $\vec{w}_0$ and then update the weight, making the step of the steepest descent of f, i.e., in the direction of the negative gradient; that is, the direction of the k-th step is computed as

$$\vec{d}_k = -\frac{\nabla f}{||\nabla f||}(\vec{w}_k), \quad (1.11)$$

and then the new value of weights is

$$\vec{w}_{k+1} = \vec{w}_k + \alpha \vec{d}_k, \quad (1.12)$$

where $\alpha \in \mathbb{R}_{\geq 0}$ is typically a small number determining the 'learning rate,' i.e., the magnitude of the step. The learning rate should be large enough to make a significant step and small enough to not step over the minimum. Fixing $f = err(\ldots)$ and composing the function in such a way to be differentiable (at least C^1 class) by selecting appropriate distance d, or its power, we can apply the above algorithm for learning the perceptron.

---

[4]For $f(x) = x^2$, the err function is called the mean square error function.

The $err(\ldots)$ function typically has local minima. Therefore, the optimization algorithm should be 'smart enough' to detect a local minimum and make an attempt to exit from it when it falls into it. Some algorithms add a stochastic element to making an optimization step (e.g., SDG — stochastic gradient descent) or add adaptive selection of step size (e.g., AdaGrad); for details, see [81].

1.6 NEURAL NETWORKS ARCHITECTURES

The most typical architectures currently are feedforward neural networks. They are created from the layers of units (perceptrons) that are stacked one over another. The first (input) layer gets the input data, and the output from this layer is forwarded to the next layers, called hidden layers, until the last (output) layer is reached. The output from the last layer is the output from the neural network. The general advantage of this architecture is the modularity. Each layer is, in general, independent[5]. Only the input size and output size must be agreed upon with surrounding layers.

A layer can be represented by the function $F_i : \mathbf{R}^{n_{i,1}} \times \mathbb{R}^{n_{i,w}} \rightarrow \mathbb{R}^{n_{i,2}}$, where $n_{i,1}$, $n_{i,2}$, and $n_{i,w}$ are the sizes of the input, output data, and the weights correspondingly. Then the action of the neural network can be mathematically represented by the functional composition:

$$F(\_, \vec{w}) := \circ_{i=1}^{L} F_i(\_, \vec{w}_i) := F_L(F_{L-1}(\ldots(F_1(\_, \vec{w}_1), \ldots), \vec{w}_{L-1}), \vec{w}_L),$$
(1.13)

where $\vec{w} = (\vec{w}_1, \ldots, \vec{w}_L)$.

The efficient training algorithm — the backpropagation algorithm — of such a network was developed in the 1990s [147] and reviewed the interest in the discipline. However, the hardware demands were not sufficient at this time. This and other unfulfilled promises of this technology started the second AI winter until the 2020s, when modern GPUs (Graphical Processing Units) were able to effectively train large neural networks of practical importance.

The backpropagation algorithm consists of two passes. The forward pass is a normal mode of ANN when the neural network computes the output $\vec{y}_i = F(\vec{x}_i, \vec{w})$. Then the error is computed.

---

[5]In general, one can construct additional data paths that omit some layers — skip links — and therefore break the direct flow of the data between layers.

It depends on the difference between output $\vec{\bar{y}}_i$ and the data label $\vec{y}_i$. Finally, layer by layer, in the opposite (backward) direction, the update of the weights $\vec{w}_i$ is propagated. There are some optimizations possible when computing gradients and appropriate activation functions are selected. Moreover, the action of each layer can be expressed as a linear algebra operation (tensor[6] multiplication) and application of an activation function component-wise, and further speed-up can be applied.

Currently, there are three main libraries that allow the construction of complicated models of feedforward neural networks. They offer a low-level interface for fast (GPU-supported) tensor operations and the backpropagationa algorithm (semi-symbolic differentiation):

- TensorFlow [1] — developed by Google company,

- PyTorch [131] — developed by Meta AI,

- JAX [24] — relatively new backend.

The high-level interface for these backends is provided by the Keras library [34]. It offers many special layers useful in various applications, like convolutional layers, recurrent layers, LSTM (Long short-term memory) layers, GRU layers, etc.

The general description of the usage of these libraries is provided in many modern practical texts on deep Learning, e.g., [145, 144, 56].

1.6.1 Other Architectures

The feedforward architectures are most common; however, there are many other topological organizations of neural network connections.

One of the oldest ones is inspired by physical models, e.g.,:

- Hopfield neural networks [66] — modeled on physical system of spins (magnetic moments) placed in the vertices of a regular lattice;

---

[6]In this context the word 'tensor' is the synonym of multidimentsional array. In mathematics the word 'tensor' is associated with a multilinear operator with specific rules of transformation when the base is changing. When one fixes the bases of all vector spaces on which the (mathematical) tensor is defined, then one obtains its representation as a multidimensional matrix — a (computer science) tensor.

- Boltzmann machines [152] — based on the physical model of spin glass;

The other architectures depend on the specific data or its representation that ANNs process, e.g.,

- graphs neural networks [173] — architectures for processing graphs;

- geometric neural networks [25] — networks that are predestined for processing geometric data;

- algebraic neural networks [129] — neural networks that employ the idea of algebraic invariance of input data.

1.7 CONCLUSIONS

Artifical neural networks shape current progress in AI research. They are applicable in various new areas. Thanks to them, the paradigm shift from the traditional direction of algorithm development to data-driven algorithms is made. In such a short summary we are unable to explain all possible directions; however, the solid foundations were summarized.

1.8 ACKNOWLEDGEMENT

The work on the chapters has been supported by the Polish National Agency for Academic Exchange Strategic Partnership Programme under Grant No. BPI/PST/2021/1/00031.

## CHAPTER 2

Applications in LLM Models and RAG Method

Łukasz Gaża

Faculty of Computer Science and Telecommunications, Cracow University of Technology, Warszawska 24, 31-155 Kraków, Poland.

Large language models (LLMs) demonstrated impressive capabilities in natural language processing tasks. They depend heavily though on pre-trained knowledge, they are not always able to access and use up-to-date information. Retrieval-augmented generation (RAG) helps to overcome this limitation by integrating external knowledge sources into the LLM's generation process. This chapter provides a comprehensive overview of RAG, exploring its underlying mechanisms and applications.

2.1 INTRODUCTION

Large language models (LLMs), such as GPT-3, revolutionized the field of natural language processing (NLP) with their ability to generate coherent and contextually relevant texts. Those models are trained on massive amounts of data, allowing them to capture patterns and relationships within language. Large language models also have some limitations: their knowledge is frozen at the time of training, limiting their ability to access and use up-to-date information.

DOI: 10.1201/9781003515302-2

Retrieval-augmented generation (RAG) seems to be a promising solution to solve this knowledge limitation. RAG combines the power of LLMs with external knowledge sources like databases, knowledge graphs, or even the internet. The additional data is used to augment the model's generation capabilities. It allows LLMs to generate more accurate and up-to-date responses.

This chapter tries to provide a comprehensive overview of LLMs and RAG, exploring their underlying mechanisms, applications, and future directions.

2.2 LARGE LANGUAGE MODELS (LLMS)

2.2.1 History of LLMs

The history of LLMs begins in the early days of AI research, where simple models based on n-gram statistics were used for tasks like text prediction and language modeling. The era of deep learning and recurrent neural networks (RNNs) in particular, because of their ability to process sequential data, allowed for the development of more sophisticated language models.

The introduction of the transformed architecture in 2016 by Vaswani et al. [165] revolutionized the field of LLMs even further. Transformers replaced RNNs with a self-attention mechanism, which allows them to process the data more efficiently and to better handle long-range dependencies. This led to the development of groundbreaking models like GPT (Generative Pre-trained Transformer) [141] and BERT (Bidirectional Encoder Representations from Transformers) [47].

The continued scaling of model size and training data has been a key driver of progress in LLMs. Models like GPT-3, with its massive 175 billion parameters, have showcased impressive capabilities in generating coherent and contextually relevant text across a wide range of domains.

Transformer Architecture

The Transformer differs significantly from traditional recurrent neural networks (RNNs) and convolutional neural networks (CNNs) in its approach to handling sequential data. The most important characteristics of the Transformer are:

1. Self-Attention Mechanism. The basis of the Transformer is the self-attention mechanism. Unlike RNNs, which process tokens sequentially, self-attention allows the model to weigh

the importance of different tokens in relation to all other to-kens in the input sequence. This is achieved by calculating attention scores for each pair of tokens, representing how much each token should "attend" to the others.

In details, the attention scores are calculated using the following steps:

- Query (Q), Key (K), Value (V) Metrics: for each token three vectors are derived: a query vector (Q), a key vector (K), and the value vector (V). They are calculated by linearly transforming the token's embedding.

- Scaled Dot-Product Attention: Attention scores are calculated as the dot product of the query vector of one token with the key vectors of all tokens, scaled by the square root of the dimension of the key vectors (to prevent vanishing gradients).

- Softmax: The attention scores are normalized using the softmax function to create a probability distribution.

- Weighted Sum: The final output for each token is a weighted sum of the value vectors of all tokens, where the weights are the attention scores.

2. Multi-Head Attention: To capture different types of dependencies, the Transformer employs multi-head attention. It performs self-attention multiple times in parallel, each time with different learned linear transformations for Q, K, and V. The outputs are concatenated and linearly transformed again to produce the final output.

3. Positional Encoding: Since self-attention does not have an inherent notion of order, positional encoding is added to the token embeddings to inject information about the relative positions of tokens in the sequence. This can be done using sinusoidal functions or learned embeddings.

4. Feedforward Network: A position-wise feedforward network is applied to each token independently. It consists of two linear transformations with a ReLU activation in between.

5. Encoder-Decoder Architecture: The Transformer typically has an encoder-decoder structure. The encoder processes

the input sequence and produces a sequence of hidden representations. The decoder generates the output sequence autoregressively, taking the encoder output and the previously generated tokens as input.

The attention mechanism can be mathematically represented as follows:

$$Attention(Q, K, V) = \text{softmax}\left(\frac{QK^T}{\sqrt{d_k}}\right) V$$

where:

- Q is the query matrix

- K is the key matrix

- V is the value matrix

- d_k is the dimension of the key vectors

The advantages of Transofrmers are:

- Parallel Processing: Unlike RNNs, Transformers can be efficiently parallelized, leading to faster training and inference.

- Global Context: Self-attention allows the model to consider all tokens in the input sequence, capturing long-range dependencies more effectively than RNNs.

- Interpretability: Attention scores provide insights into how the model is making decisions, making it more interpretable than some other architectures.

2.2.2 Mathematical Foundations of LLMs

At their core, LLMs are based on the principles of probability and statistics. They learn to predict the probability distribution of the next word in a sequence given the previous words. This is achieved through a combination of the following techniques [27].

1. Probability Theory

    - Language Modeling: LLMs are essentially probabilistic models that learn to predict the probability of the next word in a sequence given the previous words. This is based on the conditional probability:
    $$P(w_t | w_{t-1}, w_{t-2}, ..., w_1)$$

- Bayes' Theorem: LLMs can use Bayes' Theorem for inference and updating beliefs as new information is encountered:
$$P(A|B) = (P(B|A) * P(A))/P(B)$$

2. Linear Algebra

- Word Embeddings: Words are represented as vectors in a high-dimensional space, where the distance and direction between vectors capture semantic relationships.

- Matrix Operations: The core operations within transformer architectures (the basis of many LLMs) involve matrix multiplications and transformations.

3. Information Theory

- Cross-Entropy Loss: This is a common loss function used to train LLMs, measuring the difference between the predicted probability distribution and the true distribution of words.
$$H(p,q) = -\sum p(x) \log q(x)$$
where p is the true distribution and q is the predicted distribution.

4. Calculus:

- Gradient Descent: This optimization algorithm is used to adjust the parameters of the LLM during training to minimize the loss function.

- Backpropagation: This algorithm is used to calculate the gradients of the loss function with respect to each parameter, enabling efficient updates during gradient descent.

2.2.3 Applications of LLMs

Large Language Models (LLMs) have a wide range of applications, transforming how we interact with technology and information. However, they also come with certain limitations and challenges [22]. Here's a detailed look:

- Text Generation: LLMs can be used for creative writing (stories, poems, code), content creation (articles), summarization of long documents of translation. The limitations are potential for factual inaccuracies (generating text which seems to be correct but is not), difficulty controlling style across entire text and bias in generated content.

- Conversational AI: LLMs can be used for customer service chatbots, virtual assistants or educational tools. The limitations are difficulties understanding complex queries or nuanced language, lack of emotional intelligence and empathy and potential for harmful or misleading responsed.

- Question answering: LLMs can be used for information retrieval, search engines, knowledge bases, or educational platforms. The limitations may be dependence on the quality and relevance of training data, difficulty handling ambiguous or open-ended questions, and potential for biased or incomplete answers.

- Code generation: LLMs can be used for automation of repetitive coding tasks, generating code snippets from natural language descriptions or assisting in debugging. The limitations are the requirement of validation and testing of generated code, limited understanding of complex software architecture, and potential security vulnerabilities in the generated code.

- Data analysis and generation: LLMs can be used for summarizing and interpreting data trends, generating insights from large datasets, and creating visualizations from text descriptions. The limitations are the requirement of structured data input and clear prompts, limited ability to perform complex statistical analysis, and potential for misinterpreting or oversimplifying data.

2.3   RETRIEVAL-AUGMENTED GENERATION

Retrieval-Augmented Generation (RAG) has emerged as a pivotal technique for enhancing Large Language Models (LLMs) by augmenting their capabilities and addressing some of their inherent limitations. Traditional LLMs, while proficient in generating text, are often constrained by the fixed knowledge they possess from their training data. RAG addresses this by incorporating real-time access to external knowledge sources, allowing LLMs to tap into vast and up-to-date information. The concept of RAG was introduced in [90]. Since then, research in this area has flourished, with numerous techniques and architectures being proposed to enhance the retrieval and integration of external knowledge.

This integration has several profound effects. Firstly, RAG empowers LLMs to overcome knowledge limitations, enabling them to generate responses that are more factually accurate, comprehensive, and contextually relevant. This is particularly valuable in domains with rapidly evolving information, such as science, technology, or current events.

Secondly, RAG helps mitigate the problem of "hallucinations" — instances where LLMs produce plausible-sounding but factually incorrect information. By grounding their responses in retrieved evidence, RAG improves the reliability and trustworthiness of the generated text.

Moreover, RAG enhances the transparency and explainability of LLMs. By providing references to the sources used in generating responses, it allows users to verify the information's accuracy and understand the reasoning behind it. This fosters trust and accountability in AI-generated content.

The integration of RAG with LLMs gained traction in recent years due to several factors. Advancements in information retrieval techniques, such as dense passage retrieval and efficient indexing, have made it possible to retrieve relevant information from massive knowledge bases quickly and effectively. Additionally, the growing demand for accurate information, particularly in critical domains like healthcare and finance, has fueled the need for more reliable AI-generated content. Lastly, concerns about misinformation and the potential for AI to spread false narratives have emphasized the importance of transparency and explainability, both of which RAG addresses.

2.3.1 Mechanism of RAG-LLM Integration

Traditionally, LLMs are pre-trained on massive datasets, but this knowledge can quickly become outdated or insufficient for specific tasks. RAG addresses this issue by introducing a retrieval mechanism that allows the LLM to access and incorporate relevant information from external knowledge sources in real time.

The underlying principle of RAG is a two-step process. First, a retrieval model is employed to identify and select the most pertinent documents or passages from a vast knowledge base, such as Wikipedia, research papers, or even internal corporate documents. This retrieval step leverages techniques like dense passage retrieval, which maps text into a high-dimensional vector space and uses similarity measures to find the most relevant pieces of information.

Once the relevant information is retrieved, it is seamlessly integrated with the LLM to generate the final response. This integration can be achieved through various architectures, each with its unique advantages. In sequence-to-sequence models, the retrieved information and the initial query are combined and fed into the LLM as a single input. Alternatively, late interaction models allow the LLM to generate a preliminary response based on the query alone and then refine it using the retrieved information. Another approach involves neural retrievers, where the LLM itself is used to select the relevant documents, creating a more tightly coupled system.

The benefits of RAG are manifold. By grounding responses in external evidence, RAG significantly reduces the likelihood of generating factually incorrect or nonsensical outputs, a common problem known as "hallucination." Moreover, RAG enables LLMs to produce contextually relevant responses by incorporating information that is specific to the query or situation. The ability to provide references to the sources used in generating responses also enhances the transparency and explainability of the system, fostering trust and accountability.

RAG is not without its challenges. Retrieval bias, the tendency for the retrieval model to favor certain types of information, can introduce inaccuracies in the generated output. Integrating external knowledge with the LLM's internal knowledge can also be complex and requires careful calibration. Additionally, the retrieval and processing of external documents can introduce computational overhead, potentially impacting the response time.

Despite these challenges, RAG has opened up exciting new possibilities for LLMs. By dynamically accessing and integrating external knowledge, LLMs can now generate more accurate, informative, and trustworthy responses, making them valuable tools in various domains, from customer service chatbots to research assistants and content generation tools. The ongoing research in this field promises to further refine RAG techniques and expand the capabilities of LLMs, ushering in a new era of intelligent and knowledgeable language models.

2.4   LARGE LANGUAGE MODEL LANDSCAPE

The Large Language Model (LLM) landscape is a dynamic and ever-evolving arena, where a plethora of models are constantly pushing the boundaries of what's possible in natural language processing (NLP). Among the most prominent and influential players are the GPT family from OpenAI, Claude from Anthropic, the LLaMA family from Meta AI, and Gemini from Google Deep-Mind.

OpenAI's GPT models have a rich history, starting with the original GPT in 2018, which pioneered the concept of generative pre-training. This paradigm shift allowed models to learn from vast amounts of text data before being fine-tuned for specific tasks, dramatically improving their performance. GPT-2 followed in 2019, showcasing impressive text generation capabilities that sparked both excitement and concern about the potential misuse of such powerful technology. GPT-3, released in 2020, marked a significant leap in scale and performance, demonstrating the ability to generate human-like text across various domains, from creative writing to technical documentation. The subsequent releases of GPT-3.5 in 2022 [6], which underpins the popular ChatGPT, and GPT-4 in 2023, further refined the model's capabilities, adding features like improved reasoning, creativity, and safety measures.

Meanwhile, Anthropic's Claude emerged as a strong contender in 2023. Developed by a team of former OpenAI researchers with a focus on safety and ethical considerations, Claude is known for its conversational abilities and commitment to generating helpful, harmless, and honest responses. While still in limited beta access, Claude has already garnered significant attention for its potential to provide a safer and more reliable alternative to other LLMs.

The LLaMA family of models from Meta AI, also released in 2023 [161], has gained immense popularity due to its open-source nature. This has allowed researchers and developers to access and experiment with high-performing LLMs, fostering a vibrant community of innovation and accelerating progress in the field. The LLaMA family has also spawned several fine-tuned versions, such as Vicuna, Alpaca, and Koala, which have demonstrated remarkable performance on specific tasks, often rivaling or even surpassing their closed-source counterparts.

Google DeepMind's Gemini, although still not fully revealed, already has a strong position in the LLM community. Designed as a multimodal model, Gemini aims to process and understand various data types, including text, images, and potentially audio and video. While concrete details are scarce, early benchmarks suggest that Gemini could be a powerful model. Its multimodal capabilities and potential integration with Google's vast resources make it a promising contender in the evolving landscape of LLMs.

2.4.1 Comparison of Large Language Models

Comparing large language models (LLMs) involves assessing their performance across various dimensions to determine their strengths and weaknesses for specific tasks. Here's an overview of the key aspects to consider:

- Perplexity: Measures how well a model predicts a sample of text. Lower perplexity indicates better language understanding and generation capabilities.

- Accuracy: Evaluates the model's ability to produce correct answers or complete tasks successfully, especially in question-answering or text completion scenarios.

- F1 Score: Combines precision (the fraction of relevant instances among the retrieved instances) and recall (the fraction of relevant instances that were retrieved) to provide a balanced measure of a model's performance.

- ROUGE (Recall-Oriented Understudy for Gisting Evaluation): Assesses the quality of text summaries by comparing them to reference summaries.

- BLEU (Bilingual Evaluation Understudy): Evaluates the quality of machine translations by comparing them to reference translations.

- Human Evaluation: Involves human judges rating the quality, fluency, and relevance of model outputs. This can be subjective but provides valuable insights into aspects that automated metrics might miss.

Table 2.1 presents comparison of some popular LLM models and their relative performance based on various metrics [6] [161] [8].

Table 2.1 Comparison of Large Language Models

Model	Params (B)	Strengths	Weaknesses
GPT-3	175	Creative text, code generation, few-shot learning	Hallucinations, high computational cost
GPT-4	>1T (Est.)	Improved performance, stronger reasoning, enhanced safety	Less accessible, high computational cost
BERT	0.11-0.34	Powerful for NLU tasks (QA, sentiment analysis)	Not as strong in text generation
T5	0.003-11	Versatile for various NLP tasks (translation, summarization, QA)	Performance varies depending on task/data
LLaMA	7-65	Strong performance with smaller sizes, open-source	Requires substantial computational resources
Claude	-	Excellent at conversational tasks, strong reasoning, follows instructions	Less publicly available, limited metrics info
PaLM	540	Wide range of tasks (NLU, reasoning, code)	Requires substantial computational resources
Gemini	-	Multimodal (text, images, audio, etc.), strong reasoning	Details still emerging, potential for bias

2.5 FUTURE DIRECTIONS OF LLMS

The future of Large Language Models (LLMs) has promising advancements that could fundamentally change how we interact with technology and information [18]. One exciting avenue is the development of multimodal LLMs. These models will not be confined to text but will seamlessly integrate understanding and generation across different modalities like images, audio, and video. An example could be an LLM that can describe a scene in a photograph with poetic detail, answer questions about the events unfolding in a video, or even compose a musical score inspired by a written passage. Such advancements could revolutionize fields like education, entertainment, and creative arts.

Another promising direction lies in personalization. As LLMs become more sophisticated, they will be able to adapt to individual users' preferences, needs, and styles. This means that the interactions with an LLM could be tailored to the specific vocabulary, interests, and even sense of humor. This could lead to more engaging and productive interactions, whether someone is using an LLM for language learning, creative writing, or simply casual conversation.

The ethical and responsible use of LLMs is another critical area for future development. As these models become more powerful and integrated into our lives, it's essential to ensure they are used fairly and transparently. This involves addressing issues like bias in training data, ensuring that LLMs are not used to spread misinformation or manipulate public opinion, and protecting user privacy. Additionally, as LLMs become more capable of autonomous decision-making, we need to develop robust governance frameworks to ensure they are used in ways that align with human values and societal goals.

The potential for LLMs to collaborate with humans is vast and largely untapped. We can envision a future where LLMs act as cocreators, assisting writers, artists, and researchers in generating ideas, refining drafts, and exploring new creative avenues. This could lead to a renaissance of human creativity, amplified and expanded by the capabilities of AI. However, striking the right balance between human control and AI assistance will be crucial to ensure that human agency and creativity are not diminished in the process.

Finally, the development of LLMs that can continuously learn and adapt is a key goal for the future. Current models are typically trained on a static dataset and then deployed, meaning their knowledge is frozen in time. Future LLMs will need to be able to update their knowledge base in real time, incorporating new information and adapting to changes in language use and cultural context. This will ensure that LLMs remain relevant and useful in a rapidly evolving world.

2.5.1 Multimodal Large Language Models

Multimodal Large Language Models (MLLMs) represent a significant leap forward in the field of artificial intelligence [96] [155]. Unlike traditional LLMs that primarily focus on text, MLLMs are designed to process and generate information across multiple modalities, including text, images, audio, and video. This expanded capability allows them to understand and interact with the world in a more comprehensive and nuanced way, opening up a wide array of exciting applications and research directions.

At their core, MLLMs are built upon the foundation of large language models, leveraging the power of transformers and self-attention mechanisms to process and generate information. However, they extend this architecture to incorporate additional modalities, typically through specialized encoders and decoders for each modality. For instance, image encoders might use convolutional neural networks (CNNs) to extract visual features, while audio encoders might use recurrent neural networks (RNNs) to capture temporal patterns in sound.

The integration of multiple modalities allows MLLMs to learn richer representations of the world, capturing the complex interplay between different forms of information. For example, an MLLM can learn to associate the visual appearance of an object with its textual description, the sound it makes, and even its associated emotions. This deeper understanding enables MLLMs to perform a wider range of tasks than traditional LLMs, such as generating image captions, answering questions about videos, translating speech, and even creating music based on visual input.

The potential applications of MLLMs are vast and diverse. In the field of education, MLLMs could be used to create interactive learning environments that combine text, images, and videos to provide a more engaging and immersive learning experience. In healthcare, MLLMs could analyze medical images and patient records to assist with diagnosis and treatment planning. In the creative arts, MLLMs could generate new forms of art and media by combining different modalities in novel ways.

MLLMs could also revolutionize the way we interact with computers and the internet. Imagine a search engine that can understand your query not just through text but also through images or voice commands. Or a virtual assistant that can not only respond to your questions but also show you relevant images or videos.

While MLLMs hold immense promise, they also present significant challenges. One major challenge is the sheer complexity of integrating multiple modalities. Different modalities often have vastly different structures and scales, making it difficult to design models that can effectively learn from and generate information across all of them.

Another challenge is the lack of large-scale multimodal datasets. While large text datasets are readily available, curated datasets that contain aligned text, images, audio, and video are still relatively scarce. This makes it difficult to train MLLMs to their full potential.

Despite these challenges, the future of MLLMs is bright. As research progresses, we can expect to see more sophisticated models that can seamlessly integrate multiple modalities and perform a wider range of tasks. The development of new training techniques and the availability of larger and more diverse datasets will also play a crucial role in advancing the field.

2.6   CREATING CUSTOM LARGE LANGUAGE MODEL

The development of a Large Language Model (LLM) is a complex and multifaceted undertaking, demanding a substantial investment of time, computational resources, and specialized expertise. This intricate process encompasses several interconnected phases, each of which plays a crucial role in shaping the model's architecture, performance, and capabilities.

The foundation of any LLM lies in the curation and meticulous preprocessing of a massive corpus of text and code data. This extensive dataset, sourced from diverse and reputable origins such as books, scholarly articles, websites, and code repositories, serves as the raw material for the model's learning process. Rigorous preprocessing techniques, including tokenization, normalization, and meticulous filtering, are indispensable for eliminating noise, inconsistencies, and errors within the data. The quality, diversity, and sheer volume of this dataset profoundly influence the LLM's ability to generalize and perform effectively across a wide array of tasks.

Once the data is prepared, the subsequent phase involves designing the model's architecture. Transformer-based architectures, exemplified by models like GPT and BERT, have emerged as the prevailing paradigm due to their exceptional scalability and capacity to capture long-range dependencies in language. The model's size and complexity, determined by factors such as computational resources and desired performance, necessitate careful consideration. Meticulous design of the model's layers, attention mechanisms, and activation functions is pivotal for tailoring its architecture to specific tasks and domains.

The training process itself is a computationally demanding endeavor, often necessitating the utilization of high-performance hardware such as Graphics Processing Units (GPUs) or Tensor Processing Units (TPUs). The training regimen typically entails optimizing the model's parameters using gradient descent and backpropagation to minimize a loss function, such as cross-entropy loss. This iterative optimization process can span days or even weeks, depending on the model's size and the volume of training data. Throughout training, diligent monitoring of the model's performance, fine-tuning of hyperparameters, and the judicious application of techniques like early stopping are essential for achieving optimal results.

Rigorous evaluation of the trained model on a diverse set of tasks is imperative to ascertain its strengths and weaknesses. Fine-tuning the model on task-specific datasets or employing reinforcement learning from human feedback can further enhance its performance and align it more closely with human preferences. This cyclical process of evaluation and refinement is indispensable for achieving a model that generalizes well and performs reliably across various domains and applications.

Upon reaching a satisfactory level of performance, the LLM can be deployed to a production environment, such as a cloud platform or local server, to provide services to end-users. Continuous monitoring of the model's performance in real-world scenarios, coupled with the collection of user feedback and the incorporation of new data, is paramount for ensuring its ongoing effectiveness and addressing potential biases or limitations that may arise in practical applications.

Alternative strategies for LLM development include leveraging pre-trained models like GPT or BERT and fine-tuning them on domain-specific data, which can expedite the development process compared to building a model from scratch. Additionally, open-source frameworks like Hugging Face Transformers offer a wealth of tools and resources for building, training, and deploying LLMs, facilitating faster development cycles and broader accessibility.

2.6.1 Fine-tuning Existing Models

The code below shows how to fine-tune a pre-trained LLaMA model using your own dataset to make it better at a specific task. It uses the Hugging Face Transformers library to fine-tune an LLaMA model [172].

```
from transformers import AutoModelForCausalLM,
    AutoTokenizer, Trainer, %
TrainingArguments
import torch

# Load pre-trained model and tokenizer
model_name = "decapoda-research/llama-7b-hf"
tokenizer = AutoTokenizer.from_pretrained(model_name)
model = AutoModelForCausalLM.from_pretrained(model_name)

# Prepare your dataset
# (This is simplified; in practice, you'll load and
# preprocess your data here)
train_dataset = ...
eval_dataset = ...

# Training arguments
training_args = TrainingArguments(
    output_dir="./results",
    num_train_epochs=3,
    per_device_train_batch_size=16,
    per_device_eval_batch_size=64,
    warmup_steps=500,
    weight_decay=0.01,
    logging_dir="./logs",
    logging_steps=10,
    save_steps=1000,
    evaluation_strategy="steps",
    eval_steps=500,
    save_total_limit=2,
    load_best_model_at_end=True,
)
```

```
35  # Create trainer
36  trainer = Trainer(
37      model=model,
38      args=training_args,
39      train_dataset=train_dataset,
40      eval_dataset=eval_dataset,
41      data_collator=lambda data: {
42          'input_ids': torch.stack([f[0] for f in data]),
43          'attention_mask': torch.stack([f[1] for f in data]),
44          'labels': torch.stack([f[0] for f in data])
45      }
46  )
47
48  # Fine-tune the model
49  trainer.train()
```

The code does the following:

1. Initialization

   - It loads a pre-trained LLaMA model and its tokenizer. This model has already learned a vast amount about language from its initial training on a massive dataset, but it's not yet specialized for your particular task.

   - A new dataset, which is specific to this particular task (e.g., question answering, text summarization, etc.), is prepared in a suitable format.

2. Training

   - The *Trainer* class from the Hugging Face Transformers library handles the training process. It iterates through the dataset, feeding chunks of text into the LLaMA model.

   - The model predicts the next word in each sequence, and its predictions are compared to the actual next word in your dataset.

   - The difference between the model's prediction and the actual word is used to calculate a "loss" (a measure of how incorrect the model is).

   - Based on this loss, the model's internal parameters (weights and biases) are adjusted slightly using an optimization algorithm like gradient descent. This process is called backpropagation.

- This training loop continues for a specified number of epochs (passes through the entire dataset), gradually improving the model's ability to predict the next word accurately for your specific task.

Fine-tuning serves as a pivotal mechanism for enhancing the performance and adaptability of large language models (LLMs). While pre-trained on vast corpora, these models often lack the specialized knowledge required for specific tasks. Fine-tuning addresses this by exposing the model to a task-specific dataset, thereby enabling it to adapt and learn the nuances of the target domain or task.

Through the fine-tuning process, the model's internal parameters (weights and biases) are iteratively adjusted to minimize a loss function that quantifies the discrepancy between its predictions and the ground truth labels in the dataset. This adjustment facilitates a significant performance boost by optimizing the model's ability to generate relevant and accurate responses for the task at hand.

Furthermore, fine-tuning empowers practitioners to tailor the model's behavior by manipulating various hyperparameters, such as training duration and learning rate. This customization potential allows for fine-grained control over the model's performance characteristics, ensuring optimal outcomes for specific use cases.

The applications of fine-tuning are diverse and impactful. In domain adaptation scenarios, where the target task involves a specific domain like medicine, law, or finance, fine-tuning enables the model to acquire the specialized vocabulary and language patterns pertinent to that domain. Moreover, fine-tuning can be employed to induce style transfer, enabling the model to generate text in a specific style, such as formal, informal, or creative.

Fine-tuning is also instrumental in improving the model's performance on a wide array of NLP tasks. For instance, in text summarization, fine-tuning helps the model condense lengthy documents into concise summaries, while in question answering, it empowers the model to accurately respond to queries based on given information. Translation, text classification, and many other NLP tasks can also benefit from the fine-grained adaptation facilitated by fine-tuning.

2.7 SUMMARY

Large language models (LLMs) represent a paradigm shift in the field of artificial intelligence and natural language processing (NLP). Their capacity to comprehend and generate nuanced, contextually relevant text has unlocked a plethora of applications across diverse domains, from healthcare and finance to creative writing and education. The transformative potential of LLMs is evident in their ability to automate complex tasks, augment human capabilities, and democratize access to information and services.

However, the path forward is not without its challenges. Ongoing research endeavors are focused on addressing critical issues such as bias mitigation, interpretability enhancement, and resource optimization. The mitigation of inherent biases present in training data is paramount to ensuring fairness and equity in LLM-driven applications. Simultaneously, the development of techniques for interpreting and explaining LLM decision-making processes is essential for building trust and ensuring ethical deployment. Furthermore, the computational demands of training and deploying these massive models necessitate the exploration of more efficient algorithms and hardware architectures.

Despite these challenges, the future of LLMs is undeniably promising. Emerging research directions point towards the development of multimodal LLMs capable of processing and generating information across multiple modalities, including text, images, and audio. The integration of diverse modalities promises to enrich the understanding and interaction capabilities of LLMs, enabling them to tackle complex real-world tasks with greater accuracy and efficiency. Moreover, the advent of personalized LLMs tailored to individual users' preferences and needs could revolutionize user experiences across various applications, from personalized tutoring to mental health support.

As research and development efforts continue to push the boundaries of LLM capabilities, it is imperative to prioritize ethical considerations and responsible AI practices. The development of robust governance frameworks and ethical guidelines will be crucial in ensuring that LLMs are deployed in ways that align with human values and societal goals, maximizing their benefits while minimizing potential risks. With continued innovation and a commitment to ethical AI principles, the future of LLMs holds the promise of transformative advancements that will reshape the landscape of artificial intelligence and its impact on society.

## CHAPTER 3

Applications of Genetic Algorithms in Neural Networks

Wojciech Książek

Faculty of Computer Science and Telecommunications, Cracow University of Technology, Warszawska 24, 31-155 Kraków, Poland.

3.1 INTRODUCTION

Bio-inspired metaheuristics are stochastic algorithms designed to solve advanced optimisation problems. They are characterised by their ease of implementation, simple structure, and ability to avoid local minima. These algorithms often achieve significantly better results than classical optimisation methods, such as sequential quadratic programming, quasi-Newton methods, conjugate gradient methods, and fast steepest descent [125]. It should be emphasised that bio-inspired metaheuristics do not guarantee optimal solutions due to their stochastic nature, but they provide satisfactory results within a reasonable timeframe. Given that most real-world problems are NP-hard, these algorithms are widely used to tackle such challenges [2]. There are several groups of biology-inspired algorithms [5]:

- Evolutionary algorithms — based on Darwin's theory of natural selection.

DOI: 10.1201/9781003515302-3

- Swarm algorithms — based on the concept of swarm intelligence [19].

- Human-based algorithms — modelling various human activities.

- Physics-based algorithms — grounded in the laws of physics and chemistry.

Figure 3.1 illustrates the categorisation of biology-inspired algorithms and provides examples of selected algorithms. It should be noted that this represents only a small subset [68]. The scientific field related to biology-inspired algorithms is evolving rapidly. Each year, dozens of new algorithms and modifications of existing methods are developed, along with their applications in various, often highly advanced, optimisation problems. According to [142], 540 biology-inspired algorithms have been designed to date.

Biologically inspired algorithms can be divided into several main categories. Evolutionary-based algorithms include the genetic algorithm [57], differential evolution [157], evolutionary strategies [16], evolutionary programming [83], and the imperialistic competitive algorithm [13]. Swarm algorithms include particle swarm optimisation [77], the artificial bee colony algorithm [76], the bat algorithm [175], the ant lion optimiser [101], and the firefly algorithm [174]. Physical algorithms encompass methods such as simulated annealing [79], the magnetic optimisation algorithm [159], the gravitational search algorithm [146], the Archimedes optimisation algorithm [60], and the atom search algorithm [180]. The group of human-based algorithms includes the human conception algorithm [3], the brainstorm algorithm [153], student psychology-based optimisation [43], the war strategy algorithm [15], and the social-based algorithm [143].

Despite the considerable diversity among these algorithms, their underlying principles are quite similar. In the initial stage, a population is generated, with each individual (or agent) representing a potential solution to the problem. Sets of rules govern the interactions between individuals and their modifications over subsequent iterations. The algorithm concludes either after a specified number of iterations or when a solution meeting the desired accuracy is found. The effectiveness of the algorithm depends on finding a balance between exploration (searching a broad range

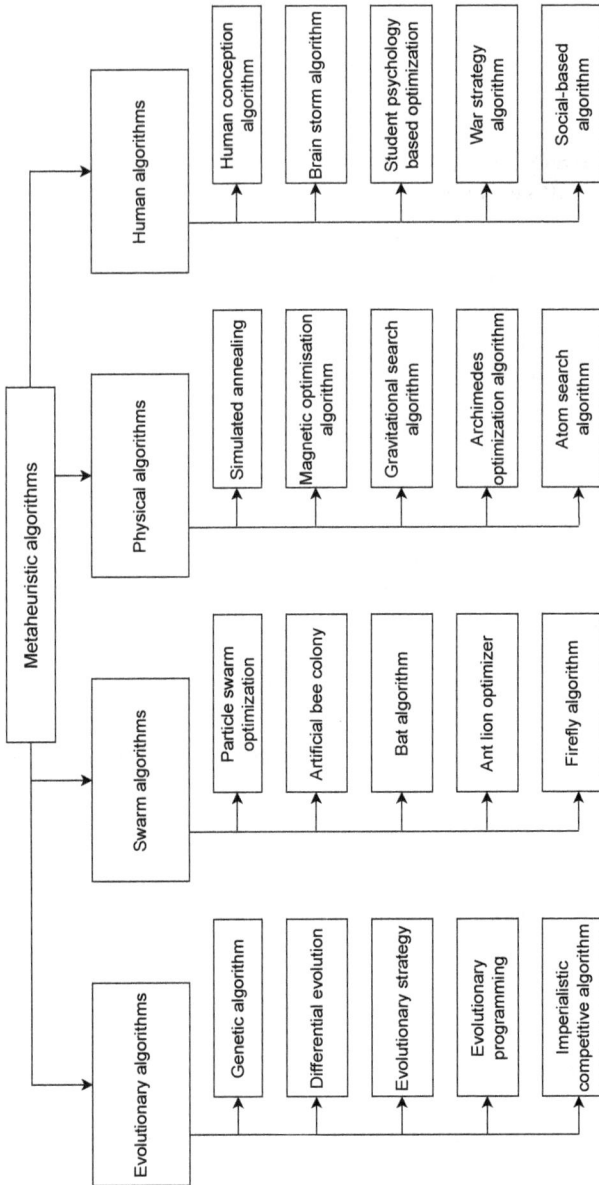

Figure 3.1 Classification of biology-inspired algorithms with selected examples [42].

of solutions) and exploitation (focusing on a promising subset of the search space) [102].

3.2 GENETIC ALGORITHM

Despite the design of new algorithms inspired by biological principles, genetic algorithms, as designed by Holland [63], remain among the most popular. These algorithms are rooted in Darwin's theory of evolution and operate on relatively straightforward principles. Initially, the algorithm starts with a randomly generated population, where each individual is encoded appropriately. This population is composed of individuals, who are made up of chromosomes, and each chromosome contains a set number of genes. Each individual represents a potential solution to the problem at hand.

A single iteration of the genetic algorithm involves selecting individuals, performing crossover operations, and applying mutations. Over successive iterations, the algorithm progressively identifies increasingly effective solutions. The operating principle of the genetic algorithm is illustrated in Figure 3.2.

The genetic algorithm operates by modelling the optimisation problem as an environment where simulated evolution occurs. Each individual in this environment represents a potential solution, and its fitness is evaluated by a fitness function, which defines our optimisation goal (whether minimisation or maximisation). Selection pressures are applied: individuals with higher fitness have a greater chance of reproducing. However, it is important not to entirely discard less fit individuals, as they may possess valuable genes crucial to solving the problem. Population diversity is maintained through processes such as crossover and mutation. Additionally, the concept of elitism is used to ensure that the best individuals are preserved despite the algorithm's inherent randomness. Typically, a fixed number of individuals is kept in the population throughout the process [35]. A critical aspect of using genetic algorithms is choosing an appropriate representation. Holland [63] proposed a binary representation, where each locus (gene position in a chromosome) can have one of two possible alleles: 0 or 1. Figure 3.3 illustrates an example of a binary individual, showing a single-point crossover and mutation.

Today, binary representation is less commonly used; real representation has become more prevalent. Figure 3.4 depicts examples of individuals using real representation. This approach

```
                    ┌──────────────┐
                    │    Begin     │
                    └──────────────┘
                           │
                           ▼
          ┌──────►┌──────────────────┐
          │       │ Initial population│
          │       └──────────────────┘
          │              │
          │              ▼
          │       ┌──────────────┐
          │       │  Selection   │
          │       └──────────────┘
          │              │
          │              ▼
          │       ┌──────────────┐
          │       │  Crossover   │
          │       └──────────────┘
          │              │
          │              ▼
          │       ┌──────────────┐
          │       │   Mutation   │
          │       └──────────────┘
          │              │
          │              ▼
     No   │         ◇ Termination ◇   Yes   ┌──────────┐
          └─────────◇  criteria  ◇─────────►│   Stop   │
                         ◇                   └──────────┘
```

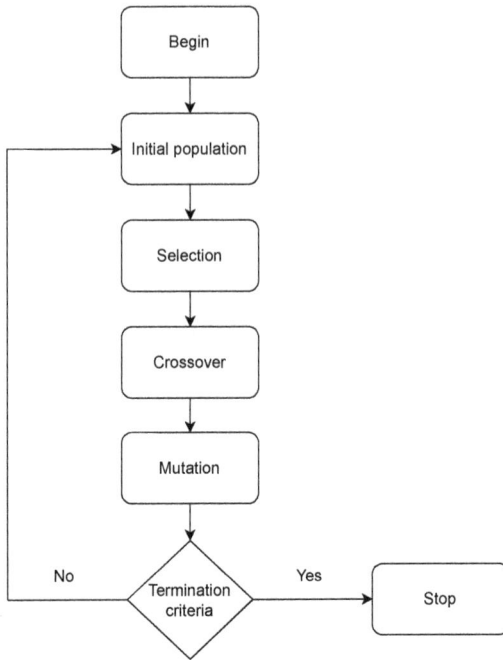

Figure 3.2 The operational principle of the genetic algorithm.

reduces the number of genes in each chromosome, resulting in lower memory usage and fewer operations during crossover and mutation processes.

The genetic algorithm requires the configuration of several parameters to function properly. Key parameters that need to be set include:

- Population size

- Number of iteration

- Fitness function

- Selection algorithm

- Crossover algorithm

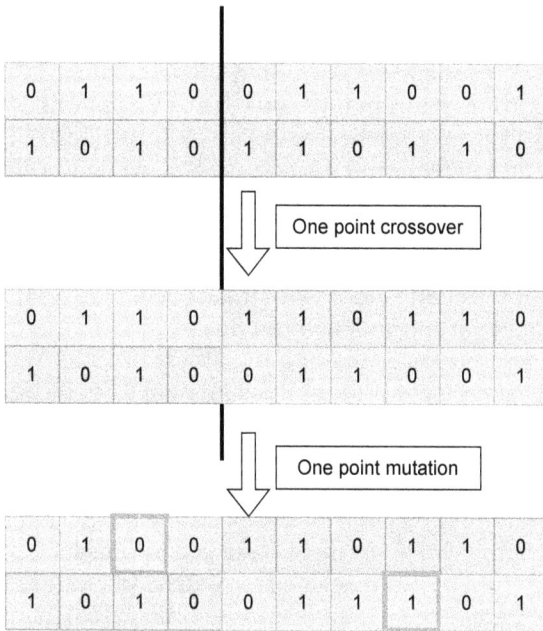

| 0 | 1 | 1 | 0 | 0 | 1 | 1 | 0 | 0 | 1 |
| 1 | 0 | 1 | 0 | 1 | 1 | 0 | 1 | 1 | 0 |

One point crossover

| 0 | 1 | 1 | 0 | 1 | 1 | 0 | 1 | 1 | 0 |
| 1 | 0 | 1 | 0 | 0 | 1 | 1 | 0 | 0 | 1 |

One point mutation

| 0 | 1 | 0 | 0 | 1 | 1 | 0 | 1 | 1 | 0 |
| 1 | 0 | 1 | 0 | 0 | 1 | 1 | 1 | 0 | 1 |

Figure 3.3 Binary representation of a chromosome.

| 0.25 | 1.44 | 2.77 | -0.13 | 10.45 | 0.33 | -4.319 | 9.12 | 3.14 | 0.001 |

| 4.55 | 1.21 | -3.53 | 2.71 | 0.11 | -0.44 | 1.89 | 7.12 | 0.31 | 5.12 |

Figure 3.4 Real representation of the chromosome.

- Mutation algorithm
- Probability of crossover
- Probability of mutation
- Elitist strategy

The selection of appropriate parameters for a genetic algorithm depends on the specific problem being addressed and the available computational resources. Various crossover algorithms are described in the literature, including single-point crossover, two-point crossover, multi-point crossover, and uniform crossover [59], as well as arithmetic crossover, simplex crossover, blend crossover, and parent-centric crossover [45]. Mutation algorithms commonly used include uniform mutation, polynomial mutation, and Gaussian mutation [31]. Selection methods also play a crucial role, with popular approaches including roulette wheel selection, tournament selection, and linear ranking selection [154]. Increasing the number of epochs and individuals can significantly impact computation time, as evaluating the fitness function for each individual is often the most time-consuming part of the algorithm. Genetic algorithms have been effectively applied to various complex optimisation problems. They have been successfully used in task scheduling [126] [179], image segmentation [7], wireless network optimisation [98], the knapsack problem [33], and mechanical engineering [20]. Recently, genetic algorithms have also been increasingly utilised in machine learning tasks, such as optimising model parameters [139], feature selection [163], and various applications related to neural networks.

3.3 THE APPLICATION OF GENETIC ALGORITHMS IN NEURAL NETWORKS

3.3.1 Optimizing Hyperparameters for Neural Networks

Genetic algorithms have proven effective in optimising machine learning model parameters, often surpassing traditional methods like grid search and random search, as well as newer techniques using libraries such as Optuna, RayTune, or HyperOpt. This approach frequently also includes simultaneous feature selection. For instance, in the study [69], genetic algorithms were used to optimise the parameters of a support vector machine (SVM). While the kernel parameter was kept constant, the parameters C and γ were optimised alongside feature selection. The research utilised well-known datasets, including German credit card data, Heart Disease (Statlog Project), and Breast Cancer (Wisconsin), and compared the results with those obtained using grid search. Genetic algorithms generally yielded better results. Similarly, in [128], genetic algorithms were employed to optimise parameters for the Nu-SVM algorithm and to perform feature selection. This study considered a broader set of parameters, including the kernel function, nu, degree, gamma, and coef0, and began with 113 features. The experiments were conducted on hyperspectral data, and the results again outperformed those achieved with grid search. In [99], genetic algorithms optimised several parameters for the XgBoost algorithm, including learning rate, n_estimators, max_depth, min_child_weight, gamma, subsample, colsample_bytree, reg_alpha, reg_lambda, scale_pos_weight, and objective. Following optimisation, the classification results for the smart grid fraud detection problem showed significant improvement. Similar experiments have been conducted with neural networks as well. In [51], the authors optimised various parameters of a standard multilayer network, including the number of hidden layers (ranging from 1 to 4), the number of neurones in each hidden layer, the activation function, solver, lambda, and alpha. These experiments were performed on well-known datasets such as Wine, Hypothyroid, Iris, and Breast Cancer, achieving classification accuracy above 90% across all datasets. Similarly, [72] explored multilayer networks with up to 5 hidden layers, optimising parameters including initial weight distribution, initial weight scale, l1 and l2 regularisation, and input dropout ratio. The datasets used in this study were Breast Cancer Wisconsin Diagnostic (BCWD), Ionosphere (Iono), Connectionist Bench — Sonar, Mines vs. Rocks (Sonar), Heart Disease (Heart), and Iris. The optimised networks achieved very high classification accuracy, exceeding 90% for 4 out of 5 datasets. For deep architectures, [85] optimised parameters for CNN and RNN networks (including LSTM and GRU), such as the number of layers, number of neurones, activation function, network optimiser, regulariser, and loss function. In [14], the focus was on optimising convolutional neural networks for the classification of crop pests, where parameters like freezing ratio, dropout rate, optimiser, and fully connected layer configuration were adjusted. The experiments covered architectures including MobileNetV2, DenseNet121, and InceptionResNetV2, leading to effective classification models. In [176], genetic algorithms were used to optimise convolutional network parameters for the MNIST digit classification problem. The parameters optimised included learning rate, dropout rates, batch size, and layer configurations, with classification results exceeding 90%. Similarly, [78] employed genetic algorithms to optimise parameters for convolutional networks in nutritional anaemia classification, focusing on training epochs, momentum, learning rate, and l2 regularisation. The best model achieved an impressive classification accuracy of 98.50%. These studies demonstrate the effectiveness of genetic algorithms in optimising neural network hyperparameters.

Based on the scikit-learn documentation [133] for the multilayer perceptron (MLPClassifier), a complex individual could be constructed to include a comprehensive set of parameters such as hidden layer sizes, neurones in hidden layers, activation function, solver, alpha, learning rate, learning rate initialisation, power t, max iterations, momentum, Nesterov's momentum, beta 1, and beta 2. Figure 3.5 illustrates example individuals representing so-

Gene number	Parameter	Value
0	Neuron in hidden layer	5
1	Activation	Logistic
2	Solver	Adam
3	Alpha	0.0001
4	Learning rate	Constant
5	Learning rate init	0.001
6	Power t	0.5
7	Max iter	500
8	Momentum	0.8
9	Nesterovs momentum	True
10	Beta 1	0.9
11	Beta 2	0.999

Gene number	Parameter	Value
0	Neuron in hidden layer	10
1	Activation	Relu
2	Solver	SGD
3	Alpha	0.0002
4	Learning rate	Invscaling
5	Learning rate init	0.002
6	Power t	0.6
7	Max iter	1000
8	Momentum	0.7
9	Nesterovs momentum	True
10	Beta 1	0.85
11	Beta 2	0.899
12	Feature 0	0
13	Feature 1	1
14	Feature n-1	0
15	Feature n	1

Figure 3.5 Hyperparameters for multilayer perceptron.

Table 3.1   Hyperparameters of the multilayer perceptron in real-valued representation

Parameter	Value	Value in Real Code	Total Range
Activation	Identity	0.0-1.0	0.0-4.0
	Logistic	1.0-2.0	
	Tanh	2.0-3.0	
	Relu	3.0-4.0	
Solver	Lbfgs	0.0-1.0	0.0-3.0
	Sgd	1.0-2.0	
	Adam	2.0-3.0	
Learning rate	Constant	0.0-1.0	0.0-3.0
	Invscaling	1.0-2.0	
	Adaptive	2.0-3.0	
Nesterovs momentum	False	0.0-0.5	0.0-1.0
	True	0.5-1.0	
Feature	Selected	0.0-0.5	0.0-1.0
	Not selected	0.5-1.0	

lutions for a multilayer perceptron with one hidden layer. As with the earlier examples in this chapter, this approach can be extended to feature selection by increasing the length of the individual representation. Here, a value of 0 indicates that a feature is excluded, while a value of 1 means the feature is included, with n representing the total number of features in the dataset. In the implementation, it is crucial to ensure that an individual with all feature values set to 0 is considered invalid (i.e., the least fit individual in the population). Similar reasoning applies to other popular libraries such as TensorFlow or PyTorch, although these libraries may involve a greater number of parameters and configurations, including more complex solvers.

The structure of the individual depicted in Figure 3.5 is appropriate. It features a mixed representation of the chromosome, incorporating integers, real numbers, text strings, and Boolean values. This mixed approach enables the application of traditional chromosome representation algorithms, such as single-point, two-point, and uniform crossover, as well as single- or two-point mutations. Alternatively, one could transition to a real-valued representation of the individual, which would facilitate the use of a broader range of crossover and mutation operators specifically designed for this format. Table 3.1 illustrates

Gene number	Parameter	Value
0	Neuron in hidden layer	5
1	Activation	1.25
2	Solver	2.21
3	Alpha	0.0001
4	Learning rate	0.35
5	Learning rate init	0.001
6	Power t	0.5
7	Max iter	500
8	Momentum	0.8
9	Nesterovs momentum	0.75
10	Beta 1	0.9
11	Beta 2	0.999

Gene number	Parameter	Value
0	Neuron in hidden layer	10
1	Activation	3.456
2	Solver	1.15
3	Alpha	0.0002
4	Learning rate	1.87
5	Learning rate init	0.002
6	Power t	0.6
7	Max iter	1000
8	Momentum	0.7
9	Nesterovs momentum	0.65
10	Beta 1	0.85
11	Beta 2	0.899
12	Feature 0	0.46
13	Feature 1	0.93
14	Feature n-1	0.22
15	Feature n	0.56

Figure 3.6 Hyperparameters of a multilayer perceptron in real-valued representation.

how individual parameters can be encoded in this real-valued representation.

In the real-valued representation, the individuals are illustrated in Figure 3.6.

This approach enables the use of standard real-valued crossover operators, such as arithmetic crossover, averaging crossover, heuristic crossover, linear crossover, and flat crossover,

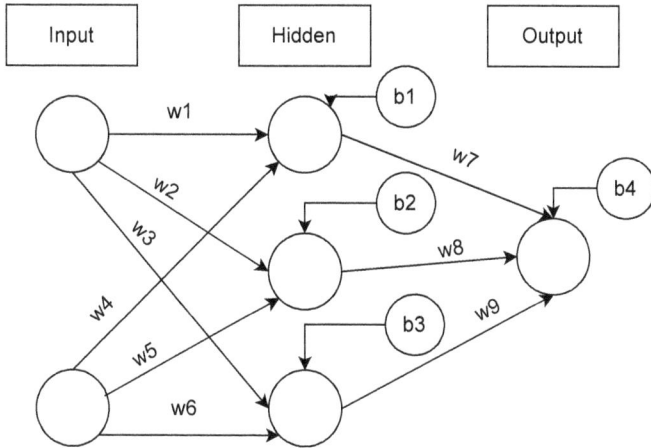

Figure 3.7   Example of a neural network.

as well as typical real-valued mutation operators like Gaussian mutation.

3.3.2   Training Neural Networks with Genetic Algorithms

The most widely used neural network training algorithm is back-propagation, introduced in 1986 [148]. Contemporary neural networks are trained using methods derived from this technique. Training a neural network involves adjusting its weights to minimise the error function. For example, consider the network depicted in Figure 3.7, which consists of three neurones in the input layer, two neurones in the hidden layer, and one neurone in the output layer.

This neural network consists of 9 weights and 4 biases. The weights and biases can be optimised using genetic algorithms, with each individual being encoded in real values and comprising 13 genes. Figure 3.8 illustrates example individuals. This method enables the use of well-established operators designed for real-valued representations.

The literature includes several instances where genetic algorithms have been employed for neural network weight selection across various problems. For example, in [178], the authors

w1	w2	w3	w4	w5	w6	w7	w8	w9	b1	b2	b3	b4
0.12	-0.33	0.21	0.51	0.71	-0.91	-0.14	0.43	0.41	0.82	-0.61	-0.28	0.01

w1	w2	w3	w4	w5	w6	w7	w8	w9	b1	b2	b3	b4
0.45	-0.27	0.21	0.94	-0.16	0.74	0.48	-0.15	0.72	0.51	-0.55	-0.91	0.59

Figure 3.8 Example of individuals for selecting neural network weights.

applied genetic algorithms to evaluate the quality of transmission in computer networks using a neural network with one hidden layer. They tested configurations with 5, 10, 15, 20, 25, 30, 40, and 50 neurones in the hidden layer, finding that the best results were achieved with 30 neurones. The achieved classification accuracy result was 95%. In [44], the authors compared genetic algorithms, the Adam optimiser, and gradient descent for training a convolutional network to detect breast cancer. The Adam optimiser achieved a classification accuracy of 85.83%, while the genetic algorithm achieved 85.49%, indicating very similar performance. The genetic algorithm was also utilised for training a neural network in the recognition of thyroid disease in [150]. The study compared the genetic algorithm with particle swarm optimisation and backpropagation, finding that the genetic algorithm yielded the best classification results on the test set. In [48], evolutionary weight selection was compared with backpropagation using the Iris dataset, showing similar results for both methods. Genetic algorithms were further applied to train advanced convolutional models for classifying croup cough in [166]. The genetic algorithm outperformed the Adam optimiser, achieving a classification accuracy of 88.32% compared to 85.83% with Adam. This review highlights that, in certain cases, genetic algorithms for training neural networks can outperform traditional gradient-based methods. The use of genetic algorithms and other biologically inspired techniques in this area has seen increased adoption in recent years.

3.3.3 Binarized Neural Networks

Neural networks have become the most widely used machine learning models for a variety of tasks in recent years. Many of these models now feature dozens or even hundreds of hidden layers. However, such deep networks often suffer from high training times and substantial memory consumption. To address these issues, researchers have been exploring efficient machine learning techniques to reduce the resource demands of complex models with billions of parameters. One promising solution is quantisation, where each neurone weight is limited to binary values (0 or 1), and the activation function is replaced with a signum function. These are known as binarised neural networks [36]. Because the parameters in such networks are not continuous, traditional gradient-based methods cannot be applied directly for training. Instead, biology-inspired algorithms such as differential evolution [122], evolutionary strategies [123], and genetic algorithms [124] offer potential alternatives. In [124], experiments were conducted on the Iris dataset, where the author tested networks with 4, 8, and 16 neurones in the hidden layer. The genetic algorithm used had the following parameters: population size of 10, elitism of 2, and 1000 iterations. Using 10-fold cross-validation, the best result—100% classification accuracy—was achieved with 16 neurones in the hidden layer. This study is preliminary and suggests the need for further research involving more datasets and more complex network architectures. The concept of binarised networks is also applicable to convolutional networks, where managing the large number of model parameters can benefit significantly from such quantisation techniques.

3.3.4 Additional Applications of Genetic Algorithms in Neural Networks

Optimising neural network hyperparameters, along with feature selection and network training, is a well-established application of genetic algorithms. However, there are several less common applications that hold potential for further scientific exploration. One such area is the use of genetic algorithms to search for optimal neural network architectures. Unlike hyperparameter optimisation or training networks with predefined weights, finding an optimal architecture is significantly more complex and requires a sophisticated approach to constructing each individual.

Research in this area includes the work of [169], where the authors introduced a novel method for encoding individuals that accommodates variable depths in convolutional networks. They tested their models on the CIFAR-10 and CIFAR-100 datasets, achieving high classification accuracies of 96.35% and 79.82%, respectively, while focusing on reducing computational time, with results obtained in just 0.2 GPU days. Another important application is the pruning of neural networks, which involves removing unnecessary branches to simplify the network and enhance performance. Genetic algorithms have been effectively employed for this purpose [140] [58] [168]. Additionally, genetic algorithms have shown promise in developing efficient ensemble models composed of convolutional networks [39] [106]. These examples illustrate how genetic algorithms can be applied to less conventional problems in neural networks, offering avenues for further research and development.

3.4 GENETIC ALGORITHMS IN PYTHON

Python is the primary language used for designing neural networks, with popular frameworks such as TensorFlow and PyTorch built on this language. When integrating genetic algorithms with neural networks, there's no need to implement these algorithms from scratch. Several Python libraries are available that can be customised to suit specific needs. These libraries offer implementations of widely used selection, crossover, and mutation methods and address parallelism concerns, which are crucial for iterative processes. Some of the most popular libraries include:

- Deap [52]
- PyGAD [54]
- PyMoo [21]

An example demonstrating the use of the PyGAD library for optimising a multivariable function. To run it, you need to install two packages: PyGAD and benchmark_functions.

- pip install pygad
- pip install benchmark_functions

The following example demonstrates the minimisation of the Ackley function with two variables, where the minimum is located at the point [0.0, 0.0] and has a value of zero. The recommended search range, as specified by the authors of the benchmark_functions [17] library, is $[-32.768, 32.768]$.

```python
from pygad import pygad
import benchmark_functions as bf

func = bf.Ackley(n_dimensions=2)

def fitness_func(ga_instance, solution, solution_idx):
    fitness = func(solution.tolist())
    return 1. / fitness

ga_instance = pygad.GA(num_generations=1000,
                       sol_per_pop=1000,
                       num_parents_mating=900,
                       num_genes=2,
                       fitness_func=fitness_func,
                       init_range_low=-32.768,
                       init_range_high=32.768,
                       gene_type=float,
                       mutation_num_genes=1,
                       parent_selection_type="tournament",
                       crossover_type="uniform",
                       keep_elitism=1,
                       K_tournament=3,
                       random_mutation_max_val=32.768,
                       random_mutation_min_val=-32.768,
                       parallel_processing=['thread', 4]
                       )

best = ga_instance.best_solution()
solution, solution_fitness, solution_idx = ga_instance.best_solution()
print("Parameters of the best solution : {solution}".format(solution=solution))
print("Fitness value of the best solution = {solution_fitness}".format(solution_fitness=1. / solution_fitness))
%\end{lstlisting}
```

The PyGAD library is particularly noteworthy, as it includes ready-made examples in Python for both classic neural networks and those built using the Keras library.

3.5 CONCLUSION

This article reviews the integration of genetic algorithms with neural networks. Genetic algorithms have been applied in several areas, including hyperparameter tuning, feature selection, network training (especially for binarised neural networks), architecture optimisation, and network pruning. Although genetic algorithms are not always the first choice for these applications, recent research has demonstrated their effectiveness and growing popularity. It is important to note that the combination of evolutionary methods with neural networks requires further investigation. As research progresses, these methods are expected to gain more traction and yield models that surpass the performance of traditional approaches.

II

Complex- and Quaternionic-valued Neural Networks and Their Applications

## CHAPTER 4

Theoretical Foundation of Complex- and Quaternionic-valued Neural Networks

Agnieszka Niemczynowicz and Radosław Kycia

Faculty of Computer Science and Telecommunications, Cracow University of Technology, Warszawska 24, 31-155 Kraków, Poland

Lluís M. García-Raffi

Universitat Politècnica de València, Cami de Vera s/n, 46022 València, Spain

4.1 INTRODUCTION

Neural networks with real-valued parameters have been widely studied and successfully applied to numerous problems. However, many types of data exhibit multidimensional or oscillatory structures that are not optimally represented in the real domain. In such cases, number systems extending the reals, such as complex numbers and quaternions, offer a more natural framework for data representation and processing.

DOI: 10.1201/9781003515302-4

The aim of this chapter is to provide the mathematical background required to build complex- and quaternionic-valued neural networks. We recall the basic algebraic properties of these number systems, introduce the corresponding neuron models, and discuss activation functions adapted to the hypercomplex setting. These foundations serve as the basis for the network architectures and applications presented in the subsequent chapters.

4.2 COMPLEX-VALUED NEURAL NETWORKS STRUCTURE

4.2.1 Basics on Complex Numbers

The history of complex numbers is a chapter in mathematics, emerging from the need to solve equations that seemed unsolvable within the framework of real numbers. The first steps towards complex numbers were taken in the 16th century. The Italian mathematician Girolamo Cardano [28], in his book entitled *Artis magnae sive de regulis algebraicis liber unus* (1545), encountered square roots of negative numbers while solving cubic and quadratic equations. Although he viewed these solutions as abstract and lacking practical significance, he acknowledged their existence. In the 17th century, mathematicians such as Rafael Bombelli began to study *imaginary* numbers more systematically. Bombelli established rules for their arithmetic, though he regarded them as purely theoretical constructs. Around the same time, René Descartes introduced the term *imaginary numbers*, highlighting their apparent lack of reality.

In the 18th century, complex numbers gained broader acceptance through the work of mathematicians such as Leonhard Euler and Jean le Rond d'Alembert. Euler introduced the symbol i to denote the square root of -1 ($i = \sqrt{-1}$) and discovered the relationship between complex numbers and trigonometry, expressed in the famous Euler's formula, namely

$$e^{i\pi} + 1 = 0.$$

Caspar Wessel and Carl Friedrich Gauss conceptualized complex numbers as points in a two-dimensional plane, with the real part as the x-coordinate and the imaginary part as the y-coordinate. This geometric interpretation significantly advanced the understanding of complex numbers.

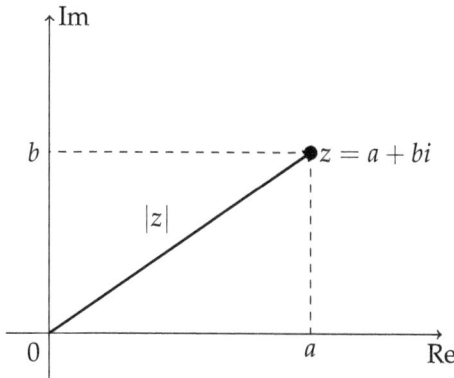

Figure 4.1   Geometric interpretation of a Complex Number.

In the 19th century, complex numbers were formally integrated into mathematical theory. Mathematicians such as Augustin-Louis Cauchy and Bernhard Riemann developed the theory of complex functions, which became a cornerstone of mathematical analysis. During this period, complex numbers found practical applications in physics, engineering, and other sciences. They became essential in fields such as electrical engineering (e.g., circuit analysis) and quantum mechanics.

Today, complex numbers are a fundamental tool in mathematics, and their applications have likely exceeded the expectations of many scientists. Complex numbers are also used in the design of neural networks due to their ability to better represent data, particularly in fields such as signal and image processing.

4.2.2   Single Complex-valued Neuron Model

In this section, we provide a brief formal description of the complex-valued neuron. Let us consider N-input single complex-valued neuron with the weights $w_n \in \mathbb{C}$, $n \in [1,\ldots,N]$ and bias $\theta \in \mathbb{C}$, where $\mathbb{C}$ is a set of complex number. For given data vector of N input signals $z_n \in \mathbb{C}$, $n \in [1,\ldots,N]$ and a complex activation function $f_C : \mathbb{C} \to \mathbb{C}$ the complex-valued output is expressed as

$$u = f_C \left(\sum_{n=1}^{N} w_n z_n + \theta \right). \tag{4.1}$$

The activation function $f_C(z)$ can be taken in a various ways.

One note have to be taken into account — in the complex domain the singularities of the functions have to be considered. The easiest way is to select analytic function, i.e., the function that locally is expressible by a convergent power series. Moreover, when $f_C(z)$ is complex differentiable then it is an analytic function in open set. The analytic function can be expressed as

$$f_C(x + iy) = u(x, y) + iv(x, y), \tag{4.2}$$

where the Cauchy conditions must be fulfilled

$$\frac{\partial u}{\partial x} = \frac{\partial v}{\partial y}, \quad \frac{\partial u}{\partial y} = -\frac{\partial v}{\partial x}. \tag{4.3}$$

This is equivalent to the condition $\frac{\partial f_C(z,\bar{z})}{\partial \bar{z}} = 0$.

One example of complex analytic functions as activation function is complex sigmoid, that is represented by the same formula as for real case:

$$\sigma_C(z) = \frac{1}{1 + exp(-z)}. \tag{4.4}$$

It possess singularities at $z = 0 \pm i(n + 1)\pi$, for $n \in \mathbb{N}$.

This situation is common to any complex analytic functions. One way to get rid of singularities is to restrict the set of possible values for weights. However, such restriction can influence the global minimum searched by back propagation algorithm. The other approach is to use non-analytic complex functions.

One of the most commonly encountered non-analytic complex activation function in the literature is the split-type activation function defined as follows

$$f_C(z) = f_R(x) + if_R(y)$$

where $z = x + iy$ is the complex-valued net input to the neuron and $f_R(u) : \mathbb{R} \rightarrow \mathbb{R}, f_R(u) = 1/1 + \exp(-u)$ is a real-valued sigmoid function. The real and imaginary components of the complex-valued output of the neuron correspond to the sigmoid functions of the real part x and imaginary part y of the net input z, respectively. Similar to real-valued neural networks, there are two main types of complex-valued neural networks: multilayer neural networks and recurrent neural networks.

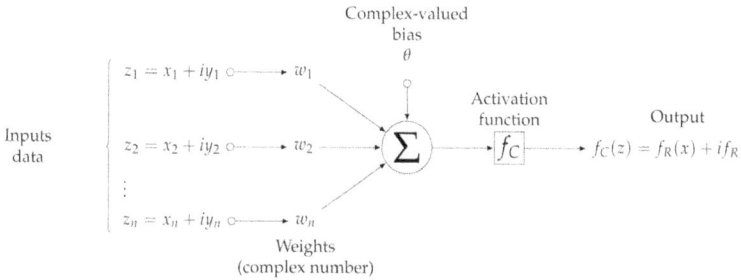

Figure 4.2 Model of a simple complex neuron.

4.2.3 Activation Functions

As it was pointed out, above, the complex analytic activation functions have that are extensions of real-valued functions have common problem with singularities. The sigmoid function was considered above. The other example is a hyperbolic function $f_C(z) = \tanh(z)$ that is extension of the real-valued functions. It has singularity for $z = 0 \pm i\frac{2n+1}{2}\pi$.

The other possible way is to use non-analytic complex functions. The common possibility is to apply real-valued activation function σ component-wise on real and imaginary part, i.e., we define $\sigma_C := \sigma_R \times \sigma_R$, which acts as

$$\sigma_C(z = x + iy = (x,y)) = (\sigma_R(x), \sigma_R(y)) = \sigma_R(x) + i\sigma_R(y). \quad (4.5)$$

In this case, application of activation function is the same as application of activation function component-wise of vectors.

The other way is to use non-analytic function that is not in the split form. One example, is to focus on radius $r = |z|$ in the polar decomposition of $z = re^{i\phi}$, e.g.,

$$\sigma_C(z) = \frac{z}{a + \frac{1}{b}|z|}, \quad (4.6)$$

where a determines the steepest of the slope, and b determines the range of σ_C.

There are generalizations of Cybenko's universal approximation theorem for complex-valued neural networks. The approximation properties strongly depend on the type of activation function [11, 10].

4.3 QUATERNION-VALUED NEURAL NETWORKS

In this section we describe neural networks that internal arithmetic is based on quaternions.

4.3.1 Basics on Quaternions

Extending complex numbers, in 1843 William Rowan Hamilton discovered multiplication table for numbers that have three imaginary units: i, j, k with multiplication table (4.7).

$$
\begin{array}{c||c|c|c|c}
 & 1 & i & j & k \\
\hline\hline
1 & 1 & i & j & k \\
\hline
i & i & -1 & k & -j \\
\hline
j & j & -k & -1 & i \\
\hline
k & k & j & -i & -1
\end{array}
\tag{4.7}
$$

For example, $q_1 = 1 + 1i$ and $q_2 = 2j$ gives

$$
q_1 q_2 = (1 + 1i)2j = 2j + 2k. \tag{4.8}
$$

They are called quaternions $\mathbb{H}$. A quaternion can be imagined as an element of $\mathbb{R}^4$ with specific multiplication.

Every, non-zero quaternion $q = a_0 + a_1 i + a_2 j + a_3 k$, where $a_i \in \mathbb{R}$ for $i \in \{0, 2, 3, 4\}$, conjugation is given by $q^* = a_0 - a_1 i - a_2 j - a_3 k$. Using this, we can define the norm of the quaternion as

$$
\|q\| = \sqrt{qq^*} = \sqrt{q^*q} = \sqrt{a_0^2 + a_1^2 + a_2^2 + a_3^2}. \tag{4.9}
$$

A non-zero quaternion q has the inverse that is given by (4.10).

$$
q^{-1} = \frac{1}{a_0^2 + a_1^2 + a_2^2 + a_3^2}(a_0 - a_1 i - a_2 j - a_3 k) = \frac{q^*}{\|q\|}. \tag{4.10}
$$

Quaternions are useful in the parametrization of rotations. We construct rotation of a vector $\vec{a} = [x, y, z] \sim xi + yj + zk$ around a vector $v = [a_1, a_2, a_3] = a_1 i + a_2 j + a_3 k$ by the angle α, we construct the 'rotor'

$$
u = \cos\left(\frac{\alpha}{2}\right) + \sin\left(\frac{\alpha}{2}\right)\frac{v}{\|v\|}. \tag{4.11}
$$

then the rotation is given by

$$
\vec{a}a \to u\vec{a}u^{-1}. \tag{4.12}
$$

One can note that both u and $-u$ define the same rotation. This is the famous double cover of rotation group by rotors. This fact is summarized by the statement that rotors form a fibre bundle over rotation group with the discrete fiber $\mathbb{Z}_2$, where $\mathbb{Z}_2$ is a group consisting of 1 and -1 with standard (real-valued) multiplication. The composition of rotations defined by the product of two rotors u_1, u_2 is given by

$$\vec{a} \to u_1 u_2 \vec{a} u_2^{-1} u_1^{-1} = u_1 u_2 \vec{a} (u_1 u_2)^{-1}, \tag{4.13}$$

where we used easily-verifiable property $u_2^{-1} u_1^{-1} = (u_1 u_2)^{-1}$.

The well-known Frobenius theorem [127] says that finite-dimensional associative division algebra over the real numbers are $\mathbb{R}$, $\mathbb{C}$, and $\mathbb{H}$. There are however other algebras that do not fulfill some of these conditions, or fulfill the other ones.

4.3.2 Quaternion-valued Neuron

The quaternions can be employed for internal arithmetic of neural networks. Starting from a single perceptron the input is encoded as a vector of quaternions, so the input dimension must be the multiplicity of 4. Then each input (a quaternion) $x_i \in \mathbb{H}$ is associated with a quaternion weight $w_i \in \mathbb{H}$, for $i \in \{1, \ldots, n+1\}$, where $x_{n+1} = 1$ and the associated bias is w_{n+1}. Moreover, take the real-valued activation function $\sigma_R : \mathbb{R} \to \mathbb{R}$. The input is a vector of 4-tules $\vec{x} = [x_1, \ldots, x_{n+1}] \in \mathbb{R}^{4n+1}$ that can be considered as a vector $\vec{x} \in \mathbb{H}^{n+1}$. The weights vector is $\vec{w} = [w_1, \ldots, w_{n+1}] \in \mathbb{H}^{n+1}$. Then the sequence of operations within the neuron are

1. Compute the product: $a = \vec{x}\vec{w} = \sum_{i=1}^{n} x_i w_i + w_{n+1} = a_0 + a_1 i + a_2 j + a_3 k$.

2. Apply σ_R componentvise, i.e., construct $\sigma_H = \times_{i=1}^{4} \sigma_R$ and then apply to a:

$$\sigma_H(a) = \sigma_R(a_0) + \sigma_R(a_1)i + \sigma_R(a_2)j + \sigma_R(a_3)k. \tag{4.14}$$

As an output of the perceptron we get a quaternion as the 4-tuple of elements.

The final function that represents the quaternion-valued perceptron is

$$y = \sigma_H \left(\sum_{i=1}^{n} x_i w_i + w_{n+1} \right). \tag{4.15}$$

The universal approximation theorem is also valid for quaternion-valued neural networks [10], i.e., quaternion-valued neural networks are universal approximates for continuous quaternion-valued functions.

4.4 CONCLUSIONS

The chapter dealt with complex and quaternionic neural networks. The general scheme for construction is evident. The scalar product of vectors with values in a given algebra is made using product within algebra. Then the component-vise activation function is applied. The result can be generalized to arbitrary algebra. The only difficult task is to prove a version of universal approximation theorem.

4.5 ACKNOWLEDGEMENT

The work on the chapters has been supported by the Polish National Agency for Academic Exchange Strategic Partnership Programme under Grant No. BPI/PST/2021/1/00031.

## CHAPTER 5

Chaos Game Representation of DNA and Hypercomplex Neural Networks Classification

Radosław Kycia, Agnieszka Niemczynowicz

Faculty of Computer Science and Telecommunications, Cracow University of Technology, Warszawska 24, 31-155 Kraków, Poland

We encode DNA strains using modified Chaos Game Representation (CGR) and then use it as an input for neural networks for multiclass classification against the species. After optimization of classical (ANN) and hypercomplex neural networks (HvNN), the results show that comparable classical neural networks have slightly better average F1 scores and smaller standard deviations. However, hypercomplex neural networks have slightly better precision and recall. The number of trainable hypercomplex neural networks is 10^4 times smaller than for classical models of

DOI: 10.1201/9781003515302-5

comparable $F1$ score. Therefore, we confirmed two hypotheses: 1) CGR encoding of DNA is sufficient to classify DNA concerning the biological species; 2) Simple architectures of hypercomplex neural networks distinguish classes similar to classical neural networks in multilabel classification in CGR-encoded species classification. However, the number of trainable parameters is about a thousand times smaller.

5.1 INTRODUCTION

Artificial Intelligence in biology is currently applied in many directions, see an overview in the articles of Nature Methods 21 (2024), e.g., [49]. Currently, the development of Large Language Models, e.g., [4], that operate on biological data can understand them [50]. Therefore, there is a need for an efficient way of encoding and processing biological data.

For DNA data, there is a need for efficient coding sequences that can be then used in learning classifiers as well as stable prediction models based on artificial neural networks.

This chapter focuses on encoding using Chaos Game Representation (CGR), a general method of graphically encoding sequences [32]. The idea is based on Dynamical Systems and was proposed in 1990 by [73]. It was inferred that this representation raises a new question about the structure of DNA. It allows us to visualize DNA structure graphically [74] and analyze species [46, 149] using distance between CGR images and clustering algorithms. One extension of this encoding is the Frequency CGR (FCGR) that counts the number of occurrences of k-mers 9 (c.f. [94]).

Since then, there have been various applications, including artificial networks, e.g., for CGR-encoded proteins [93], where comparison of Random Forrest, Support Vector Machines, and optimized fully connected ANNs with a variable number of units. The CGR encoding can provide the data that differentiates different species. The test was performed on a specific data split (no cross-validation), and optimization over the number of neurons was performed. All three methods of classification give similar results [93].

The characteristic number in encoding the DNA data is 4, which suggests using hypercomplex neural networks (HvNN) with four-dimensional hypercomplex algebras that encode 4-

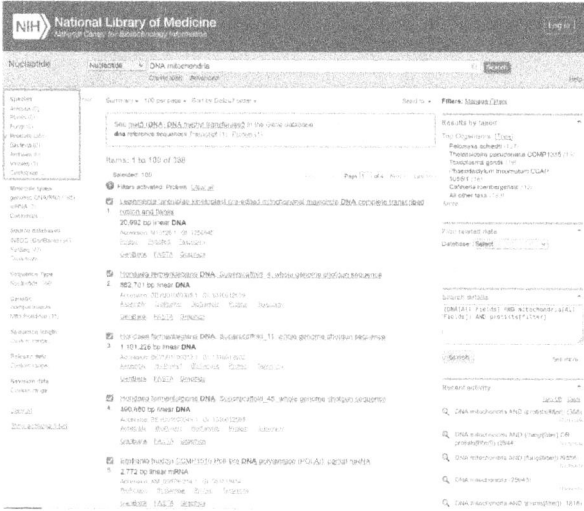

Figure 5.1 NCIB webpage with search of mitochondrial nucleotides. The red rectangle presents filters for species.

tuple as a single algebra element. The hypercomplex neural networks originate from extending the structure of neural networks.

The main aim of this preliminary study is to compare the performance of classical (ANN) and hypercomplex neural networks (HvNN) in classifying specimens using CGR-encoded DNA. For HvNN, we use *Hypercomplex Keras* package [71, 110, 111].

5.2   DATASET

For experiments, the NCBI database [107] presented in Fig. 5.1. There were selected 100 records of DNA from the following 6 specimens: Animals, Bacteria, Fungi, Plants, Protists, Viruses. Each kingdom dataset was downloaded in a FASTA format file. An example of the file for Viruses is presented in the following listing:

```
>NC_001664.4 Human betaherpesvirus 6A, variant A DNA, complete
virion genome,
isolate U1102
CGCGTTTTAAAAATTACGTCAAATCCCCCGGGGGGGCTAAAAAAAGGGGGGGTAATAACCCTAACCCTAA
CCCTAGGGCTAGCCCTAACCCTAACCCTAACCCTAGGTCTAGCCCTAACCCTAACCCTAACCCTAGG...
GCCGCTATGGGAGGCGCCGTGTTTTTCACCAACACGCGCGCCGCTGCGAGACGCGTGA

>X83413.2 Human betaherpesvirus 6A, variant A DNA, complete
virion genome,
isolate U1102
CGCGTTTTAAAAATTACGTCAAATCCCCCGGGGGGGCTAAAAAAAGGGGGGGTAATAACCCTAACCC...
```

where '. . .' indicates truncation of a large sequence of letters, and '>' is the line of comments describing the specimen. In the file, there are standard letters denoting DNA amino acids as 'A, C, T, G', however, there are letters that replace more than one letter, like 'R', which can stand for 'A' or 'G', or 'N' which can stand for all four letters in DNA sequence. We removed multi-valued symbols since they do not allow us to make CGR encoding in the later stages.

5.3 CHAOS GAME REPRESENTATION FOR DNA

Here, we present the ideas for encoding DNA using CGR. The details can be found in [73, 74, 32, 46, 149].

Since we have four letters 'A, C, T, G,' we associate with the four corners of the square as presented in Fig. 5.2.

Consider a sequence 'ACCG...' for encoding. We start from the center of the square as presented in Fig. 5.3.

Now the first letter is 'A', so we move from the center of the square to the point joining the previous position and 'A'-vertex exactly to the middle point, as presented in Fig. 5.4.

As the next letter is 'C', we move in the direction of the vertex 'C', and we place a new point in the middle point of the interval joining the previous point and the 'C'-vertex, as presented in Fig. 5.5. We repeat these jumps until we process all the letters. The final result is a CGR-encoded DNA sequence.

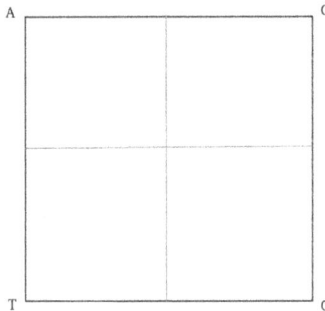

Figure 5.2 Initial configuration for CGR encoding.

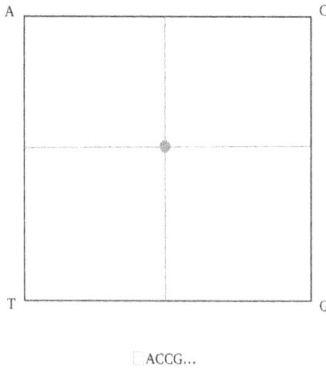

ACCG...

Figure 5.3 Starting point of encoding at the center of the square. The red rectangle and red dot represent where we are in the sequence and the rectangle, respectively.

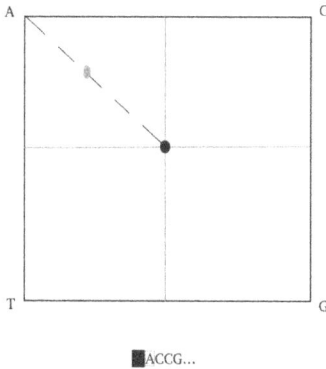

ACCG...

Figure 5.4 We move, according to the first letter 'A', to the vertex 'A' jumping to the middle point of the interval.

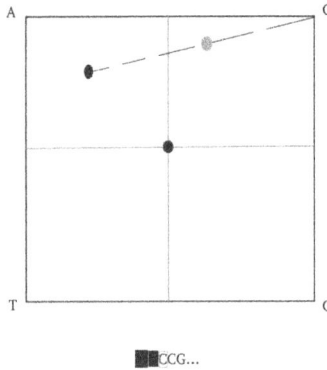

Figure 5.5 We move, according to the next letter 'C', to the vertex 'C' jumping to the middle point of the interval joining the previous stop point and 'C'-vertex.

In realizing the square, we do not have a continuum of points but a discrete raster grid; therefore, encoding depends on the raster resolution.

In the FCGR (Frequency CGR), we also note how many times each point of the grid point was visited and typically normalize these numbers to frequencies.

In our approach, we associate with each point a 4-tuple of positive integer numbers related to the number of times we visited each point with all four DNA sequence letters 'A', 'C,' 'T,' and 'G.' If we have a tuple at a given grid point, say $[5, 0, 0, 0]$, we visit this grid point with the letter A five times.

Since the viruses have relatively short DNA sequences compared to plants or animals, we restrict the processing to no more than one million letters.

In Fig. 5.6 we presented up to 1M DNA letters encoded in smaller and higher resolution. One can note that the higher resolution image has some pixels that are not visited for 1M letters. In smaller-resolution images, there are only a few unvisited pixels left out. Moreover, as noted in the paper [74], the rectangle splits in the basins of attractions of each letter.

In later processing, we used the resolution 128×128. Therefore, the initial shape of the data is $128 \times 128 \times 4$ per specimen.

Figure 5.6 Both figures represent AE006469.1 Sinorhizobium meliloti DNA sequence using 128×128 resolution (left) and 512×512 resolution (right). We associated: upper left square – 'A', upper right square – 'C', lower left square – 'T', lower right square – 'G'.

The matrix was flattened to 65536 nonnegative integers per specimen.

The classes/species were one-hot encoded into a single 6-dimensional vector with 1 representing the species, e.g., $[1, 0, 0, 0, 0, 0]$.

5.4 METHODOLOGY

In the preliminary research, we want to indicate the potential of using hypercomplex neural networks compared to classical neural networks. Therefore, we use in both cases the simplest possible architectures.

CGR encoding hides dependencies in DNA in a spatial image. These data are not coded locally within the small area of pixels for neighbor letters. Therefore, we should use dense layers to observe all data entries.

This suggests that the simplest architecture of a feedforward network has two layers:

- Input layer: Dense or Hypercomplex Dense layer; number of units will be optimized,

Table 5.1 Grid search for ANN.

Rank	Units	F1 mean	F1 Std	Trainable Parameters
1	512	83.59%	9.72%	33558022
2	256	76.89%	16.93%	16779014
3	128	55.42%	30.03%	8389510
4	64	33.62%	21.20%	4194758
5	32	20.18%	7.24%	2097382
6	16	9.96%	5.03%	1048694

- Output layer: Dense 6-unit layers with softmax activation function.

In the output layer, we use the softmax function and treat the six numerical values as the probability of belonging to each class.

Depending on the first layer type, we call the model classical (first layer is Dense) or hypercomplex (first layer is hypercomplex dense layer).

The following set of parameters was used for optimization using a standard grid search algorithm:

- Classical model: units ∈ [16, 32, 64, 128, 256, 512],

- Hypercomplex model: units ∈ [16, 32, 64, 128, 256], algebra ∈ [Quaternions, Klein4, C120, Coquaternions, Cl11, Bicomplex, Tessarines],

where Klein4 means Klein 4 group, C120 means the Clifford algebra $C\ell(2,0)$, likewise $C\ell(1,1)$ denotes the Clifford algebra $Cl(1,1)$.

Since we have multilabel classification and the number of records in each class is the same (100 specimens), we use $F1$ scoring averaged as a metric for selecting the best project. Since $F1$ metrics have no simple interpretation in multiclass classification, we also check precision and recall to check if they are close to the $F1$ score.

5.5 RESULTS

The first best ANN and HvCC architectures in the cross-validated grid search are presented in Tab. 5.1.

Likewise, we do Grid Search for the Hypercomplex model, and the results are presented in Tab. 5.2.

Table 5.2 Grid search for HvCC.

Rank	Units	Algebra	F1 mean	F1 Std	Trainable Parameters
1	256	Klein4	89.93%	4.29%	7174
2	256	Tessarines	89.48%	4.32%	7174
3	256	Bicomplex	89.24%	3.15%	7174
4	256	Cl11	88.22%	3.39%	7174
5	256	Quaternions	88.11%	4.40%	7174
6	256	Coquaternions	87.87%	2.73%	7174
7	256	Cl20	87.77%	3.48%	7174
8	128	Coquaternions	87.03%	5.16%	3590
9	128	Bicomplex	86.68%	3.71%	3590
10	128	Quaternions	85.91%	5.40%	3590
11	128	Cl20	85.73%	5.19%	3590
12	128	Cl11	85.48%	4.43%	3590
13	128	Klein4	85.19%	3.35%	3590
14	64	Cl11	84.49%	3.62%	1798
15	128	Tessarines	84.31%	3.64%	3590
16	64	Bicomplex	84.03%	4.36%	1798
17	64	Coquaternions	83.86%	4.13%	1798
18	64	Tessarines	83.19%	4.72%	1798
19	64	Quaternions	82.95%	4.33%	1798
20	32	Quaternions	82.66%	4.64%	902
21	32	Bicomplex	82.00%	4.22%	902
22	64	Cl20	81.83%	3.77%	902
23	64	Klein4	80.44%	5.98%	1798
24	32	Cl11	79.56%	3.53%	902
25	32	Tessarines	79.51%	4.86%	902
26	16	Bicomplex	79.12%	2.80%	454
27	32	Coquaternions	78.56%	5.84%	902
28	32	Cl20	77.58%	3.47%	902
29	16	Quaternions	76.35%	4.71%	454
30	16	Tessarines	75.81%	3.72%	454
31	16	Cl20	74.79%	7.33%	454
32	16	Coquaternions	74.71%	4.92%	454
33	32	Klein4	74.11%	4.45%	902
34	16	Cl11	72.68%	5.30%	454
35	16	Klein4	61.12%	7.38%	454

One can note that the HvCC models with comparable accuracy have 10^4 times less trainable parameters. Moreover, even for a small number of units, the $F1$ metric is above 50% for hypercomplex ANN.

To check if $F1$ is a good metric for multiclass classification, we compared the best classical neural network with 512 units and the hypercomplex neural network with 128 units. We get from cross-validation with $F1$, precision, and recall:

- ANN (512 units): $F1$ = 88.45% $\pm$ 6.54%, precision = 87.73% $\pm$ 6.90%, recall = 83.18% $\pm$ 13.52%;

- HvCC (128 units): $F1$ = 85.37% $\pm$ 4.87%, precision = 88.72% $\pm$ 4.30%, recall = 84.04% $\pm$ 6.31%;

Here the $F1$ score is slightly better with higher standard deviation for ANN, however precision and recall is better for HvCC model with smaller standard deviations.

5.6 CONCLUSIONS

We presented how to encode DNA sequences using the modified Chaos Game Representation method with four tuples. Then, we compared equivalent dense and hypercomplex dense neural network models. In the multiclass classification method, we found that both architectures are comparable when considering $F1$ metrics slightly in favor of the classical model. Cross-validation of models with comparable parameters have precision and recall somewhat better for hypercomplex models, which suggests better class distinction. The difference is in the total number of trainable parameters. For hypercomplex neural networks, the number of hyperparameters is about 10^4 times less. This shows the efficiency of hypercomplex neural networks.

5.7 ACKNOWLEDGEMENT

The work on the chapters has been supported by the Polish National Agency for Academic Exchange Strategic Partnership Programme under Grant No. BPI/PST/2021/1/00031.

## CHAPTER 6

Comparative Analysis of Hypercomplex and Real-Valued CNNs for Melanoma Detection: A Study on Batch Sizes

Piotr Artiemjew

Faculty of Mathematics and Computer Science, University of Warmia and Mazury in Olsztyn, ul. Słoneczna 54, 10-710 Olsztyn, Poland

This paper compares Hypercomplex Convolutional Neural Networks (HvCNNs) and classical convolutional networks (CNNs) operating on real values in the task of image classification, with a particular focus on melanoma detection. Models with different batch sizes (8, 16, 32, 64) were analysed using algebras such as quaternions, Clifford algebra ($C\ell(1,1)$), and coquaternions.

DOI: 10.1201/9781003515302-6

A special layer, `HyperConv2D`, was used to implement convolution, allowing the impact of advanced algebraic techniques on prediction quality to be assessed. The results showed that HvC-NNs, especially at smaller batch sizes, offer better stability and efficiency compared to CNNs, making them more suitable for high-precision tasks such as melanoma detection.

6.1 INTRODUCTION

Convolutional Neural Networks based on four-dimensional (4D) hypercomplex algebras (HvCNNs) are gaining increasing recognition in the field of machine learning, especially in image and signal processing tasks. Unlike classical Convolutional Neural Networks (CNNs), which operate on real numbers, these networks can use multidimensional algebraic structures, e.g., quaternions etc., to capture more complex patterns in the data. As a result, they are better able to model interdependent channels of information, such as RGB images or 3D data.

The theoretical basis for Hypercomplex-Valued Neural Networks (HvCNNs) extends from hypercomplex algebra systems, following the Cayley-Dickson procedure [26]. In fact, there are known neural network architectures based on all algebras derived from the Cayley-Dickson (modified) construction, commonly referred to in the literature as hypercomplex neural networks (cf. [9]). From the perspective of applications, new neural network architectures based on quaternions, coquaternions (split quaternions), and Clifford algebras are continuously being improved and developed as the most promising [65, 86, 87, 167].

Hypercomplex numbers, such as quaternions proposed by William Rowan Hamilton in 1843, found early applications in fields like physics and geometry. However, their broad adoption in machine learning, particularly in neural networks, became possible only with the development of modern computational resources in the late 20th century. The advancement of computing power and the rise of convolutional neural networks (CNNs) in the 1990s and 2000s provided the platform for integrating hypercomplex structures into neural networks, enabling the efficient modelling of multidimensional data [88, 84].

HvCNNs extend the capabilities of traditional real-valued CNNs by encoding data in hypercomplex forms. This allows for a more compact representation and processing of interrelated data channels, such as RGB images, 3D signals, or other multidimensional data structures. HvCNNs leverage the mathematical properties of hypercomplex numbers to reduce the number of parameters while maintaining or even improving network performance, particularly in tasks involving complex spatial relationships.

An example of HVCNN is the Quaternionic Convolutional Neural Network (QCNN), which has shown great promise in processing data like acoustic signals, medical images, and EEG data. Studies such as those by Parcollet et al. [130] and Gaudet and Maida [55] have demonstrated the efficiency of QCNNs in handling complex datasets while reducing the model size compared to real-valued CNNs.

Recent research has expanded the use of other hypercomplex algebras, such as coquaternions and the Clifford algebra $C\ell(1,1)$, to further enhance the flexibility and capabilities of HvCNNs. These advanced algebras are applied in multiplication operations within the convolutional layers, offering new opportunities in the modelling of image data and spatial signals [160]. By incorporating hypercomplex-valued convolutions, these networks can capture richer and more intricate relationships between data dimensions, making them particularly suitable for applications such as medical imaging, where precision and depth of information are crucial.

In this chapter, we will compare hypercomplex-valued convolutional networks (HvCNNs) with classical convolutional networks (CNNs) operating on real numbers in the context of image classification, specifically for melanoma detection. The analysis will involve models with different batch sizes (8, 16, 32, 64) and using various algebraic structures such as quaternions, $C\ell(1,1)$, and coquaternions. Custom layers, such as HyperConv2D, will be employed to implement hypercomplex convolutions, allowing us to investigate the impact of these advanced algebraic techniques on model performance.

We will use the PH2 dataset [100], which contains 200 dermoscopic images of skin, including cases of melanoma, atypical, and normal nevi. This dataset is commonly used to test classification and segmentation algorithms in dermatology.

6.2 METHODOLOGY

In this section, we describe the methodological approach used in our study, including both data processing and model construction. The aim is to compare the performance of Hyper-complex valued convolutional neural networks (HvCNNs) and classical convolutional networks (CNNs) in the task of melanoma detection on dermoscopic images. The steps described below include data processing, implementation of model architectures and training procedures. Particular attention is paid to the differences arising from the use of hypercomplex algebras, such as quaternions, and how to train models using data processed in different representations.

6.2.1 Data Preprocessing

The dataset used in this study comprises RGB images of skin lesions, which are preprocessed into multiple representations. Key steps include:

- **Image Resizing and Normalisation:** Images are resized to a specified dimension and normalised to scale the pixel values between 0 and 1.

- **Label Extraction:** Labels are extracted directly from image filenames to classify images into binary categories (malignant vs. non-malignant).

- **Data Augmentation:** A Keras `ImageDataGenerator` is employed to augment the data with techniques like horizontal and vertical flips to improve generalization.

- **Quaternion Representation:** Alongside the standard RGB images, quaternion-valued representations are created using HSV conversions. These representations allow for a richer encoding of image features when used in quaternion-valued convolution layers.

6.2.2 Model Architectures

The models used in this study include both real-valued CNNs and QCNNs, designed with the following configurations:

- **Real-Valued CNN:** The real-valued model consists of several convolutional layers with ReLU activation functions, followed by max-pooling layers, dropout, and a dense output layer. The model is trained using binary cross-entropy loss.

- **Quaternion-Valued CNN:** The QCNN uses a custom HyperConv2D layer, which performs convolution operations using quaternionic rules (cf. Chapter 4 in Part II). The quaternion multiplication table is defined to handle the interactions between the real and imaginary components.

- **Custom Training Procedure:** A callback function is used to monitor and store accuracy for both training and testing data after each epoch, allowing for a detailed analysis of model performance over time.

Technical details of the application of these models are in Figures 6.1 and 6.2.

6.3   EXPERIMENTAL SETUP AND ANALYSIS OF HYPERCOMPLEX AND REAL-VALUED CNNS

As part of the experimental part, we conducted a comparison of two models: the Hypercomplex CNN (HvCNN) and the Real value CNN (CNN). Both models were run on the same data splits, and we repeated each experiment five times. We used different batch size values for the experiments, namely 8, 16, 32 and 64.

In our study the code fragments available from the 'Hypercomplex-valued Convolutional Neural Networks' project [164] were applied. This project provides tools and implementations for working with networks that operate on hypercomplex domains, including the implementation of a custom layer, HyperConv2D, which enables convolutional operations with quaternions and other algebras.

6.3.1   Description of the PH2 Dermoscopic Image Dataset

The PH2 dataset [100] is a database of dermoscopic images of the skin, mainly used in studies on the automatic detection of melanoma and other melanocytic lesions. It was developed by the University of Porto and dermatologists from Pedro Hispano Hospital in Portugal, specifically for comparative research on segmentation and classification algorithms for dermoscopic images.

PH2 contains 200 dermoscopic images, of which 80 are atypical moles (nevi), 80 are common moles (common nevi) and 40 are melanoma cases. All images have a resolution of 768x560 pixels, are in 8-bit RGB format, and were standardised — obtained using

Conv2D (8 filters, 3x3, ReLU, GlorotNormal)
↓
MaxPooling2D (2x2)
↓
Conv2D (16 filters, 3x3, ReLU, GlorotNormal)
↓
MaxPooling2D (2x2)
↓
Conv2D (16 filters, 3x3, ReLU, GlorotNormal)
↓
MaxPooling2D (2x2)
↓
Conv2D (32 filters, 3x3, ReLU, GlorotNormal)
↓
MaxPooling2D (2x2)
↓
Flatten
↓
Dropout (rate=0.5)
↓
Dense (output layer)

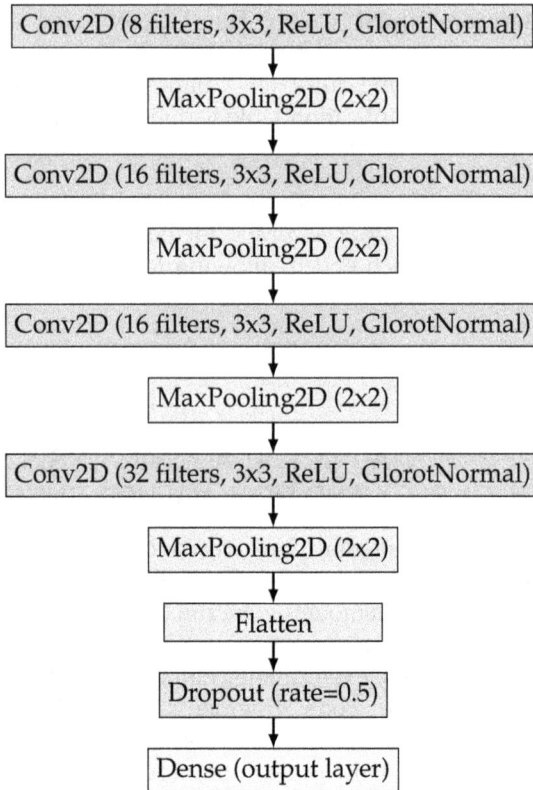

Figure 6.1 Architecture of the model **real-value CNN**. The model contains four convolution layers (Conv2D) with 8, 16, 16, and 32 filters respectively, each of size 3x3, with ReLU activation function and GlorotNormal initialisation with random seeds. Each convolutional layer is followed by a MaxPooling2D layer of size 2x2, and finally there are Flatten, Dropout (rate=0.5) and Dense layers, which are responsible for the final classification result.

20x magnification. Also available as part of the collection are manual segmentations made by dermatologists, which are provided as binary masks. These masks are used to indicate the boundaries of the skin lesion, allowing accurate comparison of the results of the segmentation algorithms. With the PH2 kit, it is possible to

```
┌─────────────────────────┐
│   Input Layer (shape)   │
└─────────────────────────┘
            ↓
┌────────────────────────────────────────────────────┐
│ HyperConv2D (2 filters, 3x3, ReLU, Selected algebra)│
└────────────────────────────────────────────────────┘
            ↓
┌─────────────────────────┐
│   MaxPooling2D (2x2)    │
└─────────────────────────┘
            ↓
┌────────────────────────────────────────────────────┐
│ HyperConv2D (4 filters, 3x3, ReLU, Selected algebra)│
└────────────────────────────────────────────────────┘
            ↓
┌─────────────────────────┐
│   MaxPooling2D (2x2)    │
└─────────────────────────┘
            ↓
┌────────────────────────────────────────────────────┐
│ HyperConv2D (4 filters, 3x3, ReLU, Selected algebra)│
└────────────────────────────────────────────────────┘
            ↓
┌─────────────────────────┐
│   MaxPooling2D (2x2)    │
└─────────────────────────┘
            ↓
┌────────────────────────────────────────────────────┐
│ HyperConv2D (8 filters, 3x3, ReLU, Selected algebra)│
└────────────────────────────────────────────────────┘
            ↓
┌─────────────────────────┐
│   MaxPooling2D (2x2)    │
└─────────────────────────┘
            ↓
┌─────────────────────────┐
│        Flatten          │
└─────────────────────────┘
            ↓
┌─────────────────────────┐
│   Dropout (rate=0.5)    │
└─────────────────────────┘
            ↓
┌─────────────────────────┐
│   Dense (output layer)  │
└─────────────────────────┘
```

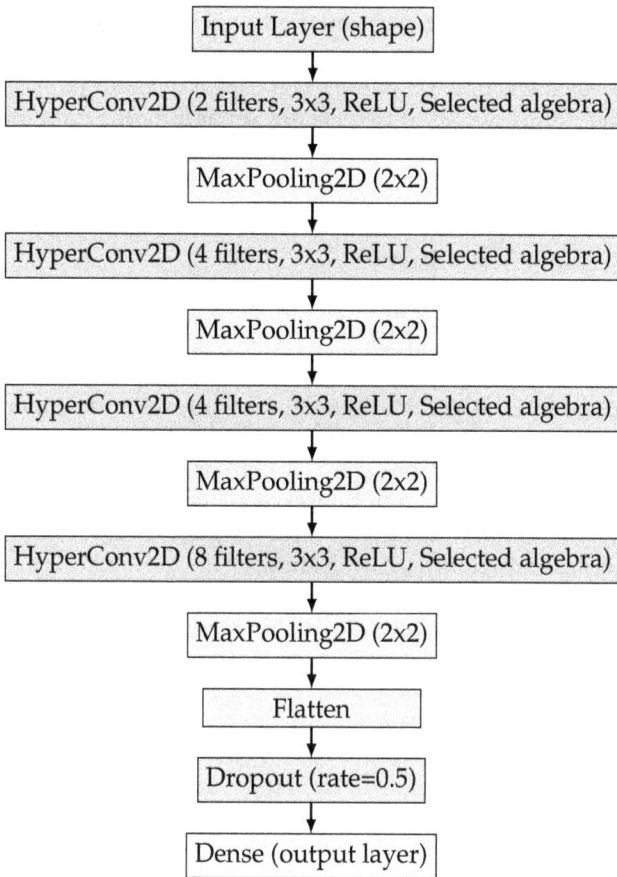

Figure 6.2 Architecture of the model **Hypercomplex-value CNN (HvCNN)**. The model consists of HyperConv2D layers with 2, 4, 4 and 8 filters of size 3x3, respectively; after each of these layers there is a MaxPooling2D layer of size 2x2. Selected algebra is presented by matrix multiplication (cf. Chapter 1 in the convolution part. At the end of the network are the Flatten, Dropout (rate=0.5) layers and the Dense layer responsible for the end result of the model.

compare different diagnostic methods, such as the segmentation and classification of skin lesions. Its images are widely used to develop systems that aid in the diagnosis of melanoma.

In our study, we used two classes separating the data into serious medical conditions vs. non-serious conditions.

6.3.2 Determining the Optimal Input Image Size for HvCNN-Q and CNN Models

From the results shown in the Figures 6.3 and 6.4, it can be seen that the performance of the HvCNN-Q and CNN models stabilises at an input image size of 200x200 pixels. For smaller image sizes (100x100), greater fluctuations in model performance are apparent, suggesting that the data are not sufficiently representative at this resolution. Conversely, for larger image sizes (300x300, 400x400, 500x500), we do not observe a significant improvement in performance, indicating that further increases in resolution do not translate into increased efficiency.

Therefore, we decided to use the 200x200 pixel size for further experiments, as it provides stable results while maintaining

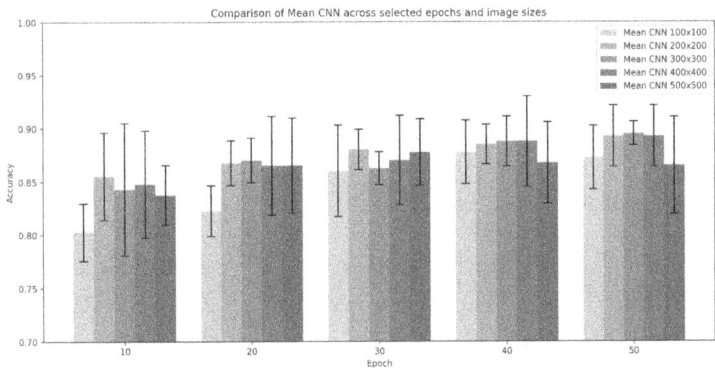

Figure 6.3 The average efficiency of the CNN network for different input image sizes (100x100, 200x200, 300x300, 400x400, 500x500) in selected learning iterations. The results show the classification efficiency for different image resolutions, with the aim of determining the optimal image size for further research.

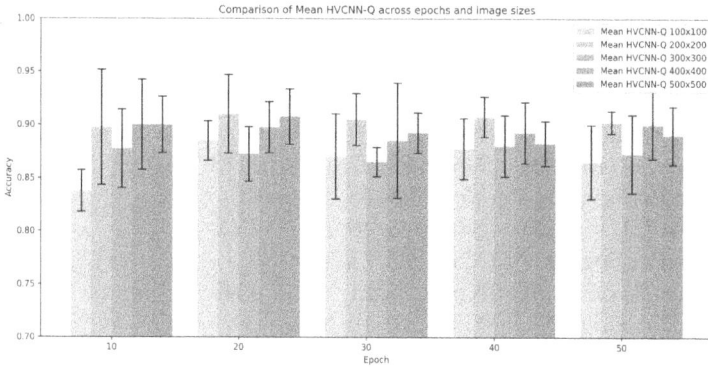

Figure 6.4 The average efficiency of the HvCNN-Q network for different input image sizes (100x100, 200x200, 300x300, 400x400, 500x500) in selected learning iterations. The results present aggregated data to help determine the optimal image size for further experiments.

moderate computational resource requirements, which is crucial for optimising the model training process.

6.3.3 Comparison of the Number of Parameters in CNNs and HvCNNs

In this section, we compare the number of parameters in two networks: a classical convolutional network (CNN) and a network based on hypercomplex values (HvCNN). Parameter counts are crucial, as they directly affect the computational complexity and training time of the models.

Parameter counting for CNN For a classical CNN network, the number of parameters for each convolutional and fully connected (Dense) layer is shown in table 6.1. In total, the number of parameters is 11553. This network consists of four convolutional layers (Conv2D), each with ReLU activation function and max-pooling layers. In addition, a Flatten layer, a Dropout layer and one Dense layer are applied at the end of the network to generate the result.

Table 6.1 Parameter counting for CNN

Layer	Output Size	Number of Parameters
Conv2D (8, 3x3)	(None, 198, 198, 8)	224
MaxPooling2D	(None, 99, 99, 8)	0
Conv2D (16, 3x3)	(None, 97, 97, 16)	1168
MaxPooling2D	(None, 48, 48, 16)	0
Conv2D (16, 3x3)	(None, 46, 46, 16)	2320
MaxPooling2D	(None, 23, 23, 16)	0
Conv2D (32, 3x3)	(None, 21, 21, 32)	4640
MaxPooling2D	(None, 10, 10, 32)	0
Flatten	(None, 3200)	0
Dropout	(None, 3200)	0
Dense (1)	(None, 1)	3201
Total		**11553**

Parameter counting for HvCNN In the case of the HvCNN, with hypercomplex selected algebras, the number of parameters is significantly less, which is due to the more efficient management of spatial information using quaternions and other algebras. As shown in table 6.2, the total number of parameters for the HvCNN is 5361, which is a much smaller value compared to the classical CNN.

Table 6.2 Parameter counting for HvCNN

Layer	Output Size	Number of Parameters
HyperConv2D (8, 3x3)	(None, 198, 198, 8)	80
MaxPooling2D	(None, 99, 99, 8)	0
HyperConv2D (16, 3x3)	(None, 97, 97, 16)	304
MaxPooling2D	(None, 48, 48, 16)	0
HyperConv2D (16, 3x3)	(None, 46, 46, 16)	592
MaxPooling2D	(None, 23, 23, 16)	0
HyperConv2D (32, 3x3)	(None, 21, 21, 32)	1184
MaxPooling2D	(None, 10, 10, 32)	0
Flatten	(None, 3200)	0
Dropout	(None, 3200)	0
Dense (1)	(None, 1)	3201
Total		**5361**

Summary   As shown in Tables 6.1 and 6.2, the HvCNN has significantly fewer parameters compared to the traditional CNN. Although HvCNN uses more complex mathematical operations such as quaternion multiplication, the reduction in the number of parameters makes the two networks more comparable in terms of computational efficiency. The computational complexity of the HvCNN is compensated by a smaller number of parameters, which can potentially improve the speed of model training while maintaining high-prediction quality.

6.4   RESULTS AND PERFORMANCE ANALYSIS

In this section, we will present the results of the performance analysis of the HvCNN-Q and CNN networks. First, we will determine the optimal input image size for the most efficient classification in the melanoma detection task on the PH2 set. Then, we will conduct a series of experiments in which, for selected values of the batch size and three algebraic structures such as quaternions, $C\ell(1,1)$ and coquaternions. We will compare which model variants best handle this practical problem. The aim is to identify the configuration that provides the best efficiency while maintaining reasonable computational complexity.

6.4.1   Performance Comparison of HvCNN and CNN vs. Batch Sizes and Hypercomplex Algebras Using Optimized Image Size

In this section, we will analyse the results of comparing the HvCNN and CNN models for different values of batch size (8, 16, 32, 64) and selected hypercomplex algebras, namely quaternions, $C\ell(1,1)$ and coquaternions. For the experiments, we will use an image size of 200x200 pixels, which has been identified as optimal in previous experiments. Our aim is to investigate which model — the HvCNN, operating in hypercomplex domain, or the classical CNN — provides better performance in classifying images from the PH2 dataset, taking into account different configurations. We will examine how changing the batch size affects the performance and stability of the models, as well as. As a result, we will identify the best model and configuration for the melanoma classification task.

Table 6.3 Comparison of Mean CNN and Mean HvCNN across selected epochs of validation (classifying TST) and batch sizes for quaternion algebra

Epoch number	Batch Size	Mean CNN	Mean HvCNN
10	8	0.8150	0.8800
10	16	0.8375	0.9000
10	32	0.8550	0.8775
10	64	0.8075	0.8375
20	8	0.8725	0.8850
20	16	0.8650	0.9075
20	32	0.8825	0.8900
20	64	0.8675	0.8825
30	8	0.8450	0.8575
30	16	0.8775	0.8925
30	32	0.8925	0.9000
30	64	0.8750	0.8825
40	8	0.8600	0.8500
40	16	0.8675	0.8825
40	32	0.8925	0.9075
40	64	0.8750	0.8950
50	8	0.8650	0.8675
50	16	0.8650	0.8900
50	32	0.8875	0.9000
50	64	0.8975	0.8950

Performance Analysis of HvCNN with Quaternion Algebra

The results (see Table 6.3 and Figure 6.5) comparing the efficiency of the HvCNN, which employs quaternion algebra for convolutional operations, with that of the classical CNN, demonstrate a clear advantage for the HvCNN in most cases. At each analysed epoch and for each batch size (8, 16, 32, 64), HvCNN achieves higher average scores compared to CNN. In particular, the HvCNN performs significantly better at lower epochs, which may indicate a faster convergence of this model. For example, as early as epoch 10 for batch size 8, the HvCNN achieves a score of 0.8800, while the CNN only achieves 0.8150. Similarly, at epoch 20 for batch size 16, the HvCNN achieves a score of

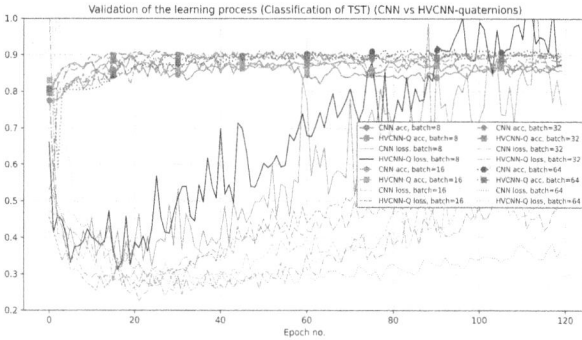

Figure 6.5 Summary of validation performance across 120 iterations for different batch sizes.

0.9075, outperforming the CNN at 0.8650. For larger batch sizes (32 and 64), HvCNN also maintains an advantage. For example, at epoch 40 for batch size 32, the HvCNN scores 0.9075, while the CNN scores 0.8925. It is also noteworthy that the HvCNN scores exhibit greater stability at larger batch sizes, whereas the CNN demonstrates more fluctuation. For example, for a batch size of 64, the CNN at epoch 10 only achieves 0.8075, while the HvCNN achieves 0.8375. The advantage of the HvCNN comes from the use of quaternions, which are more effective in modelling multivariate data. The quaternion algebra allows for more efficient processing of complex spatial relationships, which may translate into better performance in a classification task. Moreover, HvCNN shows greater stability and better performance in less time, suggesting faster model convergence compared to CNN. In summary, the use of quaternion algebra in HvCNN has clear benefits in the context of image classification. The HvCNN model outperforms the classical CNN in terms of both efficiency and stability at different batch size values, making it a more effective tool in image processing tasks.

Performance Analysis of HvCNN with Clifford Algebra

A comparison of the performance of the HvCNN network using the $C\ell(1,1)$ algebra with the classical CNN network shows clear differences in performance depending on the number of

Table 6.4 Comparison of mean CNN and mean HvCNN across selected epochs of validation (classifying TST) and batch sizes for $C\ell(1,1)$ algebra

Epoh Number	Batch Size	Mean CNN	Mean HvCNN
10	8	0.8425	0.8675
10	16	0.8225	0.8550
10	32	0.7825	0.7900
10	64	0.7950	0.8100
20	8	0.8575	0.8500
20	16	0.8900	0.8825
20	32	0.8150	0.8425
20	64	0.8000	0.8700
30	8	0.8925	0.8825
30	16	0.8850	0.9050
30	32	0.8675	0.8625
30	64	0.8400	0.8650
40	8	0.8975	0.8700
40	16	0.8750	0.9150
40	32	0.8600	0.8525
40	64	0.8825	0.8550
50	8	0.8975	0.8825
50	16	0.9100	0.9050
50	32	0.8725	0.8500
50	64	0.8750	0.8625

epochs and batch size values (see Tab. 6.4 and Fig. 6.6). In general, the HvCNN with $C\ell(1,1)$ algebra achieves a higher performance than the classical CNN, which is particularly evident at smaller batch sizes such as 8 and 16. At epoch 10 for batch size 8, the HvCNN achieves a score of 0. 8675, while the CNN achieves 0.8425. These differences persist in subsequent epochs, and for batch size 16, the HvCNN achieves a score of 0.9050 in epoch 30, which is better than the CNN, which achieves a score of 0.8850 in the same epoch. The HvCNN also shows more stable results compared to the CNN, especially for larger batch sizes such as 32 and 64. The CNN shows more fluctuation, which is especially evident at epoch 10 for batch size 64, where the CNN achieves a score of 0.7950 and the HvCNN has a score of 0.8100. The results of the HvCNN are more stable and do not show as much fluctuation,

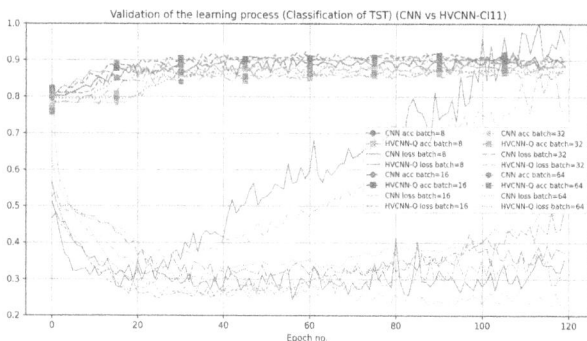

Figure 6.6 Summary of validation performance across 120 iterations for different batch sizes and $C\ell(1,1)$ algebra.

suggesting that this model is better able to cope with larger data loads. For smaller batch sizes (8 and 16), the HvCNN performs better, but the advantage over the CNN decreases at later epochs. For example, at epoch 50 for batch size 8, the two models come close to each other, where the HvCNN scores 0.8825 and the CNN scores 0.8975. For batch size 16, the HvCNN clearly outperforms the CNN, especially at epoch 40, where it scores 0.9150, while the CNN scores 0.8750. HvCNN with Cl11 algebra generally outperforms the classical CNN, especially in tasks requiring higher computational complexity and larger batch sizes. The performance of the HvCNN is more stable and the model performs better with larger data volumes. CNNs, on the other hand, exhibit greater fluctuations, especially at larger batch sizes, which may limit its effectiveness in more demanding tasks. For this reason, HvCNN with Cl11 algebra seems more suitable for tasks such as medical image classification, where stability and precision are crucial. In conclusion, HvCNN with Cl11 algebra outperforms CNN in most scenarios, offering better performance and greater stability in data processing, making this model more efficient and effective in the context of dermoscopic skin image classification tasks.

Performance Analysis of HvCNN with Coquaternion Algebra

A comparison of the performance of CNNs and HvCNNs using coquaternion algebra shows clear differences in the performance

Table 6.5 Comparison of Mean CNN and Mean HVCNN across selected epochs of validation (classifying TST) and batch sizes for coquaternion algebra

Epoh Number	Batch Size	Mean CNN	Mean HVCNN
10	8	0.8500	0.8750
10	16	0.8200	0.8275
10	32	0.7700	0.7925
10	64	0.7800	0.7950
20	8	0.8800	0.8850
20	16	0.8500	0.8625
20	32	0.8025	0.8350
20	64	0.7850	0.8150
30	8	0.8850	0.9025
30	16	0.8850	0.9000
30	32	0.8650	0.8400
30	64	0.7925	0.8350
40	8	0.8825	0.8900
40	16	0.9050	0.9025
40	32	0.8750	0.8500
40	64	0.8400	0.8550
50	8	0.8825	0.8925
50	16	0.8950	0.9100
50	32	0.9000	0.8600
50	64	0.8575	0.8600

of both models, depending on the number of epochs and batch size values (see Tab. 6.5 and Figure 6.7). In general, the HvCNN, using coquaternion algebra, shows better performance than the classical CNN, especially for smaller batch sizes such as 8 and 16. For example, at epoch 10 for batch size 8, the HvCNN achieves a score of 0.8750, while the CNN achieves 0.8500. A similar situation occurs for batch size 16 at epoch 50, where the HvCNN achieves a score of 0.9100, which exceeds the CNN score of 0.8950.

The HvCNN also shows greater stability compared to the CNN, which is particularly evident for larger batch sizes such as 32 and 64. The performance of the CNN in these cases is more variable, especially at lower epochs where the CNN scores lower.

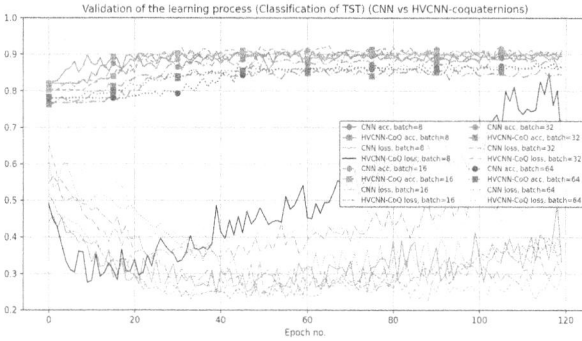

Figure 6.7 Summary of validation performance across 120 iterations for different batch sizes.

For example, at epoch 10 for batch size 32, the CNN achieves a score of 0.7700, while the HvCNN achieves 0.7925, suggesting that the HvCNN performs better when processing data at larger batch sizes.

For smaller batch sizes (8 and 16), the HvCNN consistently performs better than the CNN, especially at later epochs. At epoch 50 for batch size 8, the HvCNN scores 0.8925 and the CNN scores 0.8825. For batch size 16, the difference in favour of the HvCNN is even more pronounced. For batch size 32 and 64, on the other hand, the differences between the models are slightly smaller, but the HvCNN still shows more stable results. At epoch 50 for batch size 32, the CNN achieves a score of 0.9000 and the HvCNN 0.8600. For batch size 64, both models achieve similar results, the HvCNN with a score of 0.8600 and the CNN 0.8575.

In summary, HvCNN with the coquaternion algebra applied shows better results for most epochs and for most batch size values, especially for smaller batch sizes such as 8 and 16. For larger batch sizes, the differences become less pronounced, but HvCNN offers greater stability in results. Therefore, HvCNN with coquaternion algebra seems more suitable for tasks requiring high precision and stability, such as medical image classification. In the context of medical data processing, result stability, especially at smaller batch sizes, can be crucial, making HVCNN a more effective model for these applications.

Summary of Results for Set PH2

In experiments conducted on the PH2 dataset, CNNs and HVC-NNs were compared using different algebras (quaternion, Cl11, coquaternion) and batch size values (8, 16, 32, 64). The results show a clear advantage for HVCNN models, especially at smaller batch sizes (8 and 16). For the quaternion and coquaternion algebras, the HVCNN obtained higher results than the classical CNN, maintaining greater stability and higher efficiency at most epochs. The use of Cl11 algebras in the HVCNN also gave an advantage over the CNN, although the differences were smaller. The batch size values affected the performance of both models — smaller batch sizes showed better results in HVCNN, while larger batch sizes brought the results of both models closer, with a slight advantage for HVCNN in terms of stability.

6.5 CONCLUSIONS

Based on the experiments, it can be concluded that the HVCNN network, operating on hypercomplex values using different algebras (quaternion, Cl11, coquaternion), outperforms the classical CNN network in the task of classifying dermoscopic skin images, especially at smaller batch sizes. The quaternion and coquaternion algebras showed the greatest advantage, confirming their effectiveness in modeling complex spatial relationships. Cl11 algebra, on the other hand, offered smaller, though stable, advantages. The stability of HVCNN's results, especially at smaller batch sizes, makes this model more suitable for tasks requiring high precision, such as melanoma detection on dermoscopic images. The conclusion of this research is that HVCNNs with appropriately chosen algebras can be more efficient than classical CNNs in the context of processing complex medical images. Future research should focus on further optimisation of convolutional algorithms based on hypercomplex algebras and testing these models on other datasets to confirm their versatility and superiority in other applications.

6.6 ACKNOWLEDGEMENT

The work on the chapters has been supported by the Polish National Agency for Academic Exchange Strategic Partnership Programme under Grant No. BPI/PST/2021/1/00031.

III

Theoretical Foundation of
Computation with Neural
Networks, from Classic to Fuzzy

## CHAPTER 7

Feedforward Neural Networks as Universal Approximators

Irina Perfilieva

University of Ostrava, Institute for Research and Applications of Fuzzy Modeling, 30. dubna 22, 701 03 Ostrava, Czech Republic

Vilém Novák

University of Ostrava, Institute for Research and Applications of Fuzzy Modeling, 30. dubna 22, 701 03 Ostrava, Czech Republic

Piotr Artiemjew

Faculty of Mathematics and Computer Science, University of Warmia and Mazury in Olsztyn, ul. Słoneczna 54, 10-710 Olsztyn, Poland

This chapter focuses on the so-called model-driven approach to data analysis. This approach is closely related to the computations performed by neural networks since the latter follow a specific procedure for transforming the original data into a form well suited to solving the original, usually external, problem. The goal of this transformation is to represent the original data using feature vectors that carry sufficient information to solve the external problem. In this chapter, we propose a feature extraction

DOI: 10.1201/9781003515302-7

technique based on fuzzy modeling in general and, in particular, using the theory of *fuzzy transforms* (F-transforms). Generally speaking, features are assumed to be extracted either from raw data or from data models as a subject of training. In both cases, the process of their extraction is associated with a certain way of processing the data. In the case of neural network computing, it is determined by the network architecture; otherwise, it follows from the model chosen to represent the data.

The chapter is divided into two sections, in which we discuss two different types of data embedded in continuous or discrete spaces. In Section 7, we analyze the data as continuous functions and discuss three options for representing them in feature space: according to a Cybenko-type approximation, a Kolmogorov-type representation, and an approximate representation using F-transform theory. In Section 2 8, we deal with time series data defined in the discrete-time domain. The feature extraction method is based on the composition of two successively applied models derived from the theories of F-transforms and *fuzzy natural logic* (FNL). We demonstrate the effectiveness of the proposed method in processing long time series (about 1000 ime moments or more). The advantages of the proposed approach are as follows: (a) it provides a theoretically sound estimate of the trend cycle and its forecast, (b) its results are formulated using automatically generated natural language expressions that can also be used to explain how the results were obtained, (c) it is much less computationally expensive.

7.1   INTRODUCTION

We are living in an era of revolution in data science and machine learning, embodied in deep learning methods. Indeed, many high-dimensional learning tasks that were previously considered unachievable, such as computer vision or playing Go, etc., are in fact feasible at an appropriate computational scale. Notably, the theoretical foundation of deep learning is built on two simple algorithmic principles: first, the notion of a function representation and, second, learning via local gradient descent, typically implemented as back-propagation.

Therefore, on the one hand, we analyze function representations as realizations of the input-output relation, and on the other hand, we require that their symbolic representations can be computed according to the Artificial Neural Network (ANN) architecture. In this section, we will show how a particular representation method facilitates popular computations using neural networks and contributes to a better understanding of the meanings of the parameters involved in them.

The goal of this section is to show connections and, in some sense, similarities between the well-known classical and non-classical results on the representation and approximation of continuous functions. We focus on showing the development of classical ideas about the ways and limitations of how the latter can be represented. We examine several well-known representations in detail to demonstrate the principles and ground them in existing research, so that the reader can meaningfully apply these principles to any deep architecture he or she encounters or designs. The novelty of this section consists in introducing a new type of universal approximation of continuous functions that follows from the classical forms and extends them based on new principles that follow from partitioning the underlying domain into a set of fuzzy subsets.

This section thus focuses on analyzing some classes of symbolic realizations of input-output relations that can be computed by an ANN architecture, and then on the question of how sufficient this class is for known classes of functions.

7.2 PRELIMINARIES

We view ANN as a computational model of input-output relationships. In the case of a single-output relationship, the ANN computational model implements a set of functions characterized as follows:

$$\{f : \mathbb{R}^n \to \mathbb{R} \mid f(\mathbf{x}) = \sum_{j=1}^{N} \alpha_j \sigma(\mathbf{w}_j^T \mathbf{x} + \beta_j)\}, \qquad (7.1)$$

where $\mathbf{w}_j \in \mathbb{R}^n$, $\alpha_j, \beta_j \in \mathbb{R}$ and $\sigma : \mathbb{R} \to \mathbb{R}$ is an activation function of neurons in the hidden layer. We pose the problem: how well can a network with the above structure approximate given mappings from $\mathbb{R}^n$ to $\mathbb{R}$. Similarly, we ask about the classes of functions that can be exactly or approximately represented by the expression used in (7.1).

In the case of continuous functions, the most influential theories are associated with two papers: Kolmogorov's representation theorem [82] and Cybenko's approximation theorem [40]. We will first focus on the paper [40] because the analytical representation proposed there corresponds, up to notation, to the representation given above in (7.1). This form is repeated in most works related to neural network computing, see for example [67, 70], since it represents all the calculations performed.

The author [40] proved that the set of functions defined on $[0,1]^n$, $n \geq 1$ and represented in the form (7.1) is dense in $C([0,1]^n)$ with respect to the supremum norm if σ is a sigmoid function[1]. It follows that an ANN as a feedforward neural network with one hidden layer and a finite number of neurons computes any continuous function following the form (7.1) with any chosen accuracy.

The result proven in [40] and being a theoretical platform for ANN computational models does not provide an algorithm for setting the parameters used in (7.1). The latter can be either manually selected based on a sufficiently large set of samples (see, for example, [95]), or calculated based on the training results using various approaches developed in machine learning.

Below we give an idea (borrowed from [95]) of how an ANN (called a "shallow neural network") can be designed to compute a univariate continuous function over a bounded real interval.

7.3 APPROXIMATION OF A CONTINUOUS FUNCTION ACCORDING TO CYBENKO USING A "SHALLOW NEURAL NETWORK"

In this section, we show how one can design a "shallow" feedforward neural network with one hidden layer to compute (approximately) any continuous function given enough of its arguments and corresponding values. Let us assume that a neural network is implemented using a certain device that computes an output signal for a given input signal. Following the computation method, the formal relationship between the input and the output can be described by a function presented in (7.1). The latter will be identified with the function computed by the neural network.

Therefore, to design a feedforward neural network with one hidden layer, we must define the parameters that we see in (7.1). These are: the internal $\mathbf{w}_j \in \mathbb{R}^n$ and external weights $\alpha_j \in \mathbb{R}$, the biases $\beta_j \in \mathbb{R}$, the activation function $\sigma : \mathbb{R} \to \mathbb{R}$ and the

---

[1]In fact, this statement was proved in [40] for any continuous discriminatory function σ, of which the sigmoid function is a special case.

number of terms N. A widely used method for finding ANN parameters is *machine learning*. It is used when there is no additional information about the data source except for a set of input-output pairs. On the other hand, when information about the data source is known to some extent, the ANN parameters can be at least partially specified at the initialization stage.

Below we will show that, assuming that the input-output pairs belong to a continuous function, we can propose an initialization in which an ANN designed in this way not only computes the function values at known inputs, but also predicts the output values at intermediate points with good accuracy. To do this we rely on the two above explained facts: (i) an ANN with one hidden layer computes outputs following (7.1); (ii) any continuous function can be approximated by a function represented using (7.1) with a reasonable accuracy.

In what follows, we assume that a univariate continuous function f is known at any point in the interval $[a, b]$. In order to propose the ANN initialization, which leads to a reasonable approximation, we use the method (see [95]) based on the approximate representation of a continuous function in the form of a piecewise constant function.

In more detail, we partition $[a, b]$ into N subintervals using nodes $x_1, \dots, x_N$ where $x_1 = a$, $x_N = b$, $x_{i+1} - x_i = \frac{b-a}{N-1}$, and first approximate f by the function represented by

$$f(x) \approx \sum_{i=1}^{N-1} f(x_i)\chi_{[x_i, x_{i+1})}, \tag{7.2}$$

where $\chi_{[x_i, x_{i+1})}$ is the characteristic function of the interval $[x_i, x_{i+1})$ defined on $[a, b]$. Thus, to build a neural network that approximately computes a continuous function, it is sufficient to build as many of its constant parts as necessary. We call each constant part represented by $f(x_i)\chi_{[x_i, x_{i+1})}$ a *tower function*.

Below we give an example[2] of an ANN with two neurons in the hidden layer which calculates the following tower function on $[-1, 1]$.

$$f_{[0.2, 0.3]}(x) = \begin{cases} 0.6, & \text{if } 0.2 \le x \le 0.3, \\ 0, & \text{otherwise.} \end{cases}$$

---

[2]https://www.mathematik.uni-wuerzburg.de/fileadmin/10040900/2019/
Seminar-Artificial-Neural-Network-24-9.pdf

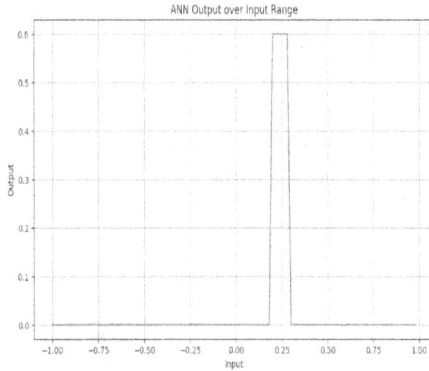

Figure 7.1 Tower function.

Since we have identified the function computed by the neural network with the function represented as (7.1), we must find parameters in (7.1) that guarantee the correct computation of $f_{[0.2,0.3]}$. First, we choose the sigmoid activation function

$$\sigma(x) = \frac{1}{1 + \exp(-x)},$$

and the number of terms $N = 2$. Then the representation of $f_{[0.2,0.3]}$ with (7.1) can be as follows:

$$f_{[0.2,0.3]}(x) = 0.6\sigma(1000x - 200) + (-0.6)\sigma(1000x - 300). \quad (7.3)$$

Thus, the remaining parameters of the neural network that computes $f_{[0.2,0.3]}$ are as follows: internal weights $w_1 = w_2 = 1000$, external weights $\alpha_1 = 0.6$, $\alpha_1 = -0.6$, biases $\beta_1 = -200$, $\beta_2 = -300$. In Figure 7.1 we show the plot of the actual function computed by the ANN according to the prescription given in (7.3).

Using this simple example, we can design an ANN with one hidden layer to compute (approximately) any continuous function given enough of its arguments and corresponding values. If f is such a function defined on $[a, b]$, then we first represent it using (7.2). We then see that each term in (7.2) has the form $f(x_i)\chi_{[x_i, x_{i+1})}$, which coincides with the tower-like function, and hence can be computed by the ANN identified by

$$f_{[x_i, x_{i+1}]}(x) = f(x_i)\sigma(1000x - x_i) + (-f(x_i))\sigma(1000x - x_{i+1}).$$
$$(7.4)$$

Since the above ANN computing the tower-like function requires two hidden neurons, the new ANN computing the function f on $[a, b]$ (as a sum of corresponding towers) requires $2N - 2$ hidden neurons. It follows that the larger N, the better the approximation of f can be achieved by the function computed by the ANN.

We illustrate this conclusion in Figure 7.2, where we show the plots of two actual functions computed by ANNs with $N = 20$ and $N = 200$ hidden neurons, respectively. Both computed func-

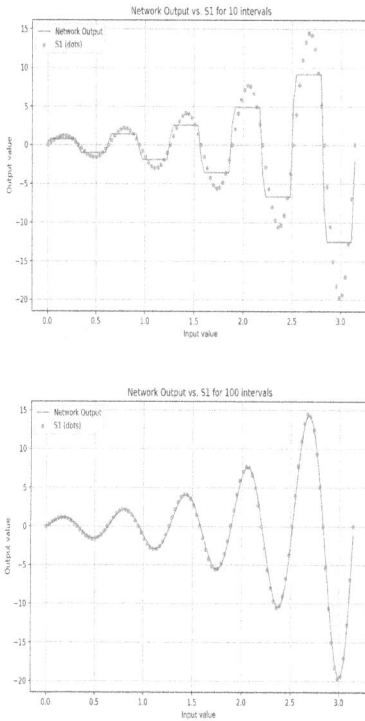

Figure 7.2 Two plots of the actual functions computed by an ANN with $N = 20$ (left) and $N = 200$ (right) hidden neurons, respectively, that approximate the function $f(x) = e^{\frac{\pi x}{100}} \sin\left(\frac{\pi x}{10}\right)$ on $[0, \pi]$.

tions approximate the following one

$$f(x) = e^{\frac{\pi x}{100}} \sin\left(\frac{\pi x}{10}\right), x \in [0, \pi]. \tag{7.5}$$

The parameters of both ANNs were chosen based on two piecewise constant approximations (7.2) of the function (7.5) based on two sets of its argument-value pairs (as input and output data values), which consist of 10 and 100 pairs, respectively.

7.4 APPROXIMATION OF A CONTINUOUS FUNCTION ACCORDING TO KOLMOGOROV-ARNOLD REPRESENTATION

In contrast to the paper [40], the representation of a multivariate continuous function in a bounded domain using addition and univariate functions (known as the "Kolmogorov-Arnold representation") was presented in [82] constructively, albeit as a convergent sequence of functions. This representation is used in an ANN with a different architecture [92] than the one discussed above in Section 7.3. This new approach promises significant improvements in accuracy and interpretability over traditional multilayer perceptrons (MLPs). The main difference with MLPs is that KANs have learnable activation functions that are assigned to the network "edges". In other words, KANs have no vector-weights every weight parameter is replaced by a univariate function.

Kolmogorov-Arnold Networks (KANs) are currently being intensively analyzed by followers [138, 92, 37]. Below we repeat more details of the original Kolmogorov proof, following [103]. These details will be used in the new ANN architecture proposed in this section.

7.5 KOLMOGOROV–ARNOLD THEOREM AS AN ANSWER TO HILBERT'S 13TH PROBLEM

Theorem 1 (Kolmogorov, Arnold, Kahane, Lorentz, Sprecher). For any $n \in \mathbb{N}$, $n \geq 2$, there exist real numbers $\lambda_1, \ldots, \lambda_n$ and continuous functions $\phi_k : \mathbb{I} \to \mathbb{R}$, $k = 1, \ldots, 2n + 1$, where $\mathbb{I} = [0, 1]$, with the property that for every continuous function $f : \mathbb{I}^n \to \mathbb{R}$ there exists a continuous function $g : \mathbb{R} \to \mathbb{R}$ such that for each

$(x_1, \ldots, x_n) \in \mathbb{I}^n$,

$$f(x_1, \ldots, x_n) = \sum_{k=1}^{2n+1} g(\lambda_1 \phi_k(x_1) + \cdots + \lambda_n \phi_k(x_n)). \qquad (7.6)$$

The cited above theorem (see it in e.g., [12, 82, 103]) is an answer to Hilbert's 13th problem, in which he conjectured that there is a continuous function $f : \mathbb{I}^3 \to \mathbb{R}$, where $\mathbb{I} = [0,1]$, which cannot be expressed in terms of composition and addition of continuous functions from $\mathbb{R}^2 \to \mathbb{R}$, that is, as composition and addition of continuous real-valued functions of two variables. It took more than 50 years to prove Hilbert's conjecture false.

7.5.1 Kolmogorov superposition theorem - necessary details

In this section, we present one of the generalizations of Kolmogorov superposition theorem and outline its proof proposed by the Swedish mathematician Torbjörn Hedberg, which uses the work of George Gunter Lorentz, Jean-Pierre Kahane and David A. Sprecher.

Below, we reproduce some necessary details of the proof for the $n = 2$ variable case taken from [103]. Precisely, the proof will be given for a continuous function $f : \mathbb{I}^2 \to \mathbb{R}$ and the following representation formula:

$$f(x_1, x_2) = \sum_{k=1}^{5} g(\phi_k(x_1) + \lambda \phi_k(x_2)), \qquad (7.7)$$

where only the function g depends on f, in contrast to λ and ϕ_k.

We recall that the set $C(\mathbb{I})$ of all continuous functions from $\mathbb{I}$ into $\mathbb{R}$ is a complete metric space and it is a Banach space with the norm $\|f\| = \sup_{x \in [0,1]} f(x)$. Denote $\Phi_k(x_1, x_2) = \phi_k(x_1) + \lambda \phi_k(x_2)$ and rewrite (7.7) as follows:

$$f(x_1, x_2) = \sum_{k=1}^{5} g(\Phi_k(x_1, x_2)). \qquad (7.8)$$

Lemma 1. There exists a real number λ such that for any $x_1, x_2, y_1, y_2 \in \mathbb{Q}^3$,

$$x_1 + \lambda y_1 = x_2 + \lambda y_2, \Rightarrow x_1 = x_2, y_1 = y_2.$$

---

[3]$\mathbb{Q}$ denotes the set of rational numbers

The following lemma [103] gives a constructive (albeit preliminary) description of the functional parameters, which, after substituting these parameters into the right-hand side of the equality (7.7), makes it approximate.

Lemma 2. We fix λ satisfying Lemma 1 and choose a function $f \in C(\mathbb{I}^2)$ such that $\|f\| = 1$. We define a set $U_f \subseteq [C(\mathbb{I})]^5$ such that $(\phi_1, \ldots, \phi_5) \in U_f$ if and only if there exists $g \in C(\mathbb{R})$, such that

$$|g(t)| \leq \frac{1}{7} \text{ for } t \in \mathbb{R}, \tag{7.9}$$

and

$$\left| f(x,y) - \sum_{k=1}^{5} g(\phi_k(x) + \lambda\phi_k(y)) \right| < \frac{7}{8} \text{ for } (x,y) \in \mathbb{I}^2. \tag{7.10}$$

Then U_f is an open dense subset of $[C(\mathbb{I}^5)]$.

Below we repeat (following [103]) all necessary details of the proof.

7.6 COVERING OF THE SET $\mathbb{I}$

Let us start with the describing special functions $(\phi_1, \ldots, \phi_5) \in [C(\mathbb{I})]^5$.

1. First, we will choose a sufficiently large $N \in \mathbb{N}$, which we will refine later.

2. We then create a *covering* of $\mathbb{I}$ using five sets $\mathcal{P}_1, \ldots, \mathcal{P}_5$ of subintervals:

$$\mathbb{I} = \bigcup_{k=1}^{5} \mathcal{P}_k, \tag{7.11}$$

where $\mathcal{P}_k$ consists of all subintervals of $\mathbb{I}$ that remain after all intervals $[\frac{s}{N}, \frac{s+1}{N}]$ with $0 \leq s < N$, $s \equiv k - 1 \pmod 5$ are deleted. These remaining N_k intervals, which we will consider closed, will be called *intervals of rank k*.

3. For each $k \in \{1, \ldots, 5\}$, we define $\phi_k : \mathbb{I} \to \mathbb{R}$, as any continuous function satisfying the following requirements:

   - ϕ_k is a constant equal to a rational number on each interval of rank k in $\mathcal{P}_k$;

- $\phi_k(x) \neq \phi_k(y)$ for x and y in different intervals of rank k;
- $\phi_k(x) \neq \phi_j(z)$ for x in any interval of rank k and z in any interval of rank j, $k \neq j$.

4. The set of all quintuples $(\phi_1, \ldots, \phi_5)$ of functions satisfying the conditions, given in steps 1-3, will be denoted by U_N. It is clear that the union $\bigcup_{n>N} U_n$ is an open dense subset of $[C(\mathbb{I})]^5$.

In Figure 7.3 [103], we show subintervals that belong to the set $\bigcup_{k=1}^5 \mathcal{P}_k$.

Figure 7.3 Subintervals that belong to the set $\bigcup_{k=1}^5 \mathcal{P}_k$. The top line corresponds to intervals of the rank $k = 1$.

7.7 COVERING OF THE SET $\mathbb{I}^2$

Let $\mathbb{I}$ be covered by subintervals according to (7.11), such that for each $k = 1, \ldots, 5$, $\mathcal{P}_k$ consists of the intervals of rank k. The length of any interval of rank k does not exceed the value of $\frac{4}{N}$. The Cartesian product of two intervals of rank k (one lying in $\{0 \leq x \leq 1\}$ and one lying in $\{0 \leq y \leq 1\}$, will be called a *rectangle of rank k*. Rectangles of rank k will be denoted by $R_{k,1}, R_{k,2}, \ldots$. The (finite) union of all rectangles of all ranks k, $k = 1, \ldots, 5$ forms a *covering* of $\mathbb{I}^2$.

We observe that the squared Euclidean distance between any two points (x, y) and (x', y') in any rectangle of rank k is less than (or equal to) $\frac{32}{N^2}$.

Continuing with the description of functions that fulfill (7.9) and (7.10), we define

$$\Phi_k(x, y) = \phi_k(x) + \lambda\phi_k(y), \ k = 1, \ldots, 5, \quad (7.12)$$

where $\Phi_k : \mathbb{I}^2 \to \mathbb{R}$. By the properties of functions ϕ_k, we have that each function Φ_k is constant on each rectangle $R_{k,j}$, $j = 1, \ldots$, of rank k, and by Lemma 1, the (constant) value of Φ_k on $R_{k,j}$, denoted by $\Phi_{k,j}$, does not equal to the (constant) value $\Phi_{k',j'}$ of $\Phi_{k'}$ on $R_{k',j'}$, if $k \neq k'$ and $j \neq j'$.

Now to prove (7.9) and (7.10) we choose $N_f \in \mathbb{N}$ such that

$$|f(x, y) - f(x', y')| < \frac{1}{56} \text{ if } (x - x')^2 + (y - y')^2 \leq \frac{32}{N_f^2}. \quad (7.13)$$

Let us substitute the value N_f for N at step 1) of the procedure for describing special functions $(\phi_1, \ldots, \phi_5) \in [C(\mathbb{I})]^5$.

Finally, we define $g : \mathbb{R} \to \mathbb{R}$ on the set $\{\Phi_{k,j} \mid k = 1, \ldots, 5; j = 1, \ldots\}$:

$$g(\Phi_{k,j}) = \begin{cases} \frac{1}{7}, & \text{if } f(x, y) > 0 \text{ for all } x, y \in R_{k,j}, \\ -\frac{1}{7}, & \text{if } f(x, y) < 0 \text{ for all } x, y \in R_{k,j}, \end{cases} \quad (7.14)$$

and extend g to the whole $\mathbb{R}$ in a piecewise-linear fashion so that $|g(t)| \leq \frac{1}{7}$ for all $t \in \mathbb{R}$. Obviously, (7.9) is satisfied. Moreover, it can be proved (see e.g.,[103]) that (7.10) is satisfied as well.

It is important to remark that (7.14) defines function g on a finite set of points, i.e., on the set $\{\Phi_{k,j} \mid k = 1, \ldots, 5; j = 1, \ldots\}$ using only two values: $\frac{1}{7}$ and $-\frac{1}{7}$.

7.7.1 Eliminating the Dependence of Φ_k on f

The next step in proving the validity of the representation (7.7) (in its equivalent form (7.8)) is to confirm inequalities similar to (7.9) and (7.10), but in which the functions $\phi_1, \ldots, \phi_5$ do not depend on f. For this purpose, a technique based on the Baire category theorem [4] is used. The following lemma is proved in [103]:

----

[4]*The Baire category theorem, see [104].* Let (X, d) be a complete metric space. If $X_1, X_2, \ldots$ is a sequence of open dense subsets of X, then the set $\bigcap_{n=1}^{\infty} X_n$ is also dense in X.

Lemma 3. Let λ satisfy Lemma 1. There exist functions $\phi_1, \ldots, \phi_5 \in C(\mathbb{I})$, such that given $f \in C(\mathbb{I}^2)$ there exists $g \in C(\mathbb{R})$, such that

$$|g(t)| \leq \frac{1}{7}\|f\|, \, t \in \mathbb{R}, \qquad (7.15)$$

and

$$\left\| f - \sum_{k=1}^{5} g \circ \Phi_k \right\| < \frac{8}{9}\|f\|, \qquad (7.16)$$

where $\Phi_k(x, y) = \phi_k(x) + \lambda\phi_k(y)$, $k = 1, \ldots, 5$.

To prove Lemma 3, we again assume that $\|f\| = 1$ and (following [103]) choose a sequence of functions $h_1, h_2, \ldots$ from $C(\mathbb{I}^2)$ (from the unit sphere of $C(\mathbb{I}^2)$) such that the set $\{h_j : j \in \mathbb{N}\}$ is dense in the unit sphere of $C(\mathbb{I}^2)$. Therefore, there exists $m \in \mathbb{N}$ such that $\|f - h_m\| \leq \frac{1}{72}$.

By the assertion in Section 7.5.1, each function h_j defines the set $U_{h_j} \subseteq (C(\mathbb{I}))^5$. Since each such U_{h_j} is a dense open subset of the complete metric space $(C(\mathbb{I}))^5$, then by the Baire category theorem their intersection is non-empty. Denote this intersection as V and choose $(\phi_1, \ldots, \phi_5) \in V$. For the chosen function h_m and for $(\phi_1, \ldots, \phi_5) \in V \subseteq U_{h_m}$ there exists a continuous function g, $\|g\| \leq \frac{1}{7}$ such that

$$\left\| h_m - \sum_{k=1}^{5} g \circ \Phi_k \right\| < \frac{7}{8}.$$

Therefore,

$$\left\| f - \sum_{k=1}^{5} g \circ \Phi_k \right\| \leq \|f - h_m\| + \left\| h_m - \sum_{k=1}^{5} g \circ \Phi_k \right\| < \frac{1}{72} + \frac{7}{8} = \frac{8}{9}.$$

7.8 COMPLETING THE PROOF OF KOLMOGOROV SUPERPOSITION THEOREM

Having all the necessary technical details, we can formulate the main statement (for the case of $n = 2$ variables), known as Kolmogorov superposition theorem [82]. For the original proofs, we refer to [12, 82, 103].

Theorem 2 (A. N. Kolmogorov). There exist a real number λ and continuous functions $\phi_k : \mathbb{I} \to \mathbb{R}$, $k = 1, \ldots, 5$, where $\mathbb{I} = [0, 1]$, which have the property that for every continuous function $f : \mathbb{I}^2 \to \mathbb{R}$ there is a continuous function $g : \mathbb{R} \to \mathbb{R}$ such that for all $(x_1, x_2) \in \mathbb{I}^2$,

$$f(x_1, x_2) = \sum_{k=1}^{5} g(\phi_k(x_1) + \lambda\phi_k(x_2)). \tag{7.17}$$

Proof. By Lemma 3, there exists a real number λ and functions $\phi_1, \ldots, \phi_5 \in C(\mathbb{I})$ such that for a given $f \in C(\mathbb{I}^2)$ there exists $g \in C(\mathbb{R})$ such that the inequalities (7.15) and (7.16) hold.

Let us put $f_0 = f$ and $g_0 = g$. Then, define recursively the sequences of functions $f_1, f_2, \ldots \in C(\mathbb{I}^2)$ and $g_1, g_2, \ldots \in C(\mathbb{R})$ as follows. First, define

$$f_{j+1} = f_j - \sum_{k=1}^{5} g_j \circ \Phi_k,$$

and then, using Lemma 3, find g_{j+1} such that the inequalities (7.15) and (7.16) are satisfied after replacing g to g_{j+1}, and f to f_{j+1}, $j = 0, 1, \ldots$.

By Lemma 3,

$$\|g_j\| \leq \frac{1}{7}\|f_j\|,$$

and

$$\|f_{j+1}\| = \left\| f_j - \sum_{k=1}^{5} g_j \circ \Phi_k \right\| < \frac{8}{9}\|f_j\|.$$

Thus,

$$\|f_{j+1}\| < \frac{8}{9}\|f_j\| < \cdots < \left(\frac{8}{9}\right)^{j+1}\|f_0\| = \left(\frac{8}{9}\right)^{j+1}\|f\|, j = 0, 1, \ldots$$

Hence,

$$\|g_{j+1}\| \leq \frac{1}{7}\|f_{j+1}\| < \frac{1}{7}\left(\frac{8}{9}\right)^{j+1}\|f\|,$$

and the series $\sum_{j=0}^{\infty} g_j$ converges in norm to a $g \in C(\mathbb{R})$. Therefore, we have

$$f = \sum_{j=0}^{\infty} (f_j - f_{j+1}) = \sum_{j=0}^{\infty} \sum_{k=1}^{5} g_j \circ \Phi_k = \sum_{k=1}^{5} g \circ \Phi_k.$$

This completes the proof of the theorem. □

7.9 INVERSE F-TRANSFORM APPROXIMATION IN-SPIRED BY KOLMOGOROV'S SUPERPOSITION THEOREM

The purpose of this section is to combine two areas of research: classical functional analysis and a part of fuzzy modeling based on the functional representation of fuzzy sets with $[0,1]$-valued membership functions[5]. In particular, we aim to show that a constructive proof of Lemma 2 can be given in a functional space with a fuzzy partition, using analytical form known as the inverse fuzzy transform [136].

The F-transform (originally, *fuzzy transform*) [136] is a particular integral transform whose peculiarity consists in using a *fuzzy partition* of a universe of discourse (usually, $\mathbb{R}$). We observe that the F-transform method was motivated by the ideas and techniques of fuzzy logic (see, e.g., [80, 108, 177]) and especially by the Takagi-Sugeno models [158]. In addition, the idea of a fuzzy partition was derived from observing a collection of antecedents in a fuzzy rule based system. The direct F-transform components are possible consequents in the Takagi-Sugeno model with singletons.

The F-transform has two phases: direct and inverse (see details in [136, 135]). The direct F-transform is applied to functions from $L_2(\mathbb{R})$ and maps them linearly onto sequences (originally finite) of numeric/functional components. Each component is a weighted orthogonal projection of a given function on a certain linear subspace of $L_2(\mathbb{R})$. Weights are determined by fuzzy sets in a corresponding fuzzy partition. The inverse F-transform is applied to a sequence of components and transforms it linearly into a function from $L_2(\mathbb{R})$ that, in general, is different from an original one. The inverse F-transform smoothly approximates the original function.

In terms of the two representations (exact and approximate) discussed above, we aim to show that the inverse F-transform takes advantage of both of them. More details will be given after we introduce the inverse F-transform.

---

[5]A fuzzy set A in $\mathbb{R}$ is identified by its $[0,1]$-valued membership function $A : \mathbb{R} \to [0,1]$.

7.9.1 Fuzzy Partition

The notion of a fuzzy partition does not have a nonambiguous meaning in fuzzy literature. We will not go into full detail but concentrate on an evolution of this notion in connection with the F-transform (see [64, 134, 156]).

A *fuzzy partition with the Ruspini condition* was introduced in [136] as a collection of bell-shaped fuzzy sets $A_1, \ldots, A_n$ on the real interval $[a, b]$ with continuous membership functions, such that for all $x \in [a, b]$,

$$\sum_{k=1}^{n} A_k(x) = 1.$$

This partition can be characterized as a "partition-of-unity".

A fuzzy partition with the generalized Ruspini condition was introduced in [156]. The generalization consists in replacing the "partition-of-unity" by a "fuzzy r-partition", $r \geq 2$, where for all $x \in [a, b]$, the above given condition changes to

$$\sum_{k=-r+2}^{n+r-1} A_k(x) = r.$$

This type of partition was investigated in [156, 62], where the focus was on smoothing or filtering data using the inverse F-transform.

In [134], a generalized fuzzy partition without the Ruspini condition was proposed with the purpose of obtaining a better approximation by the inverse F-transform.

Below, in Definition 1, we introduce a particular case of a generalized fuzzy partition that is determined by a generating function. We say that function $a : \mathbb{R} \to [0, 1]$ is a *generating function of a fuzzy partition* (shortly, a generating function), if it is non-negative, continuous, even, bell-shaped and moreover, it vanishes outside $[-1, 1]$ and fulfills $\int_{-1}^{1} a(t) \, dt = 1$. Below, we give two examples of a generating function, which we call the *raised cosine* and *triangle*, respectively:

$$a^{cos}(t) = \begin{cases} \frac{1}{2}(1 + \cos(\pi t)), & -1 \leq t \leq 1, \\ 10, & \text{otherwise.} \end{cases} \tag{7.18}$$

$$a^{tr}(t) = \begin{cases} 1 - |t|, & -1 \leq t \leq 1, \\ 0, & \text{otherwise.} \end{cases} \tag{7.19}$$

Generating function a produces infinitely many *rescaled* functions $a_H : \mathbb{R} \to [0,1]$ such that

$$a_H(t) \overset{\text{def}}{=} a\left(\frac{t}{H}\right),$$

where H is a positive number called a *scale factor*.

Definition 1. Let $a : \mathbb{R} \to [0,1]$ be a generating function of a fuzzy partition, i.e., a is non-negative, continuous, even, bell-shaped, vanishes outside $[-1,1]$ and fulfills $\int_{-1}^{1} a(t)\, dt = 1$. Let, moreover, $h > 0$, $t_k = t_0 + k \cdot h$, $k \in \mathbb{Z}$, be uniformly distributed nodes[6] in $\mathbb{R}$. Let $H > \frac{h}{2}$ and a_H be an H-rescaled version of a. With each node t_k, we correspond the translation $a_k(t) = a_H(t_k - t)$. We say that the set $\{a_k, k \in \mathbb{Z}\}$ establishes an (h, H)-uniform a-generated fuzzy partition of $\mathbb{R}$. Functions a_k are called basic functions.

By the condition $H > \frac{h}{2}$, each point from $\mathbb{R}$ is "*covered*" by at least one basic function - by this we mean that the value of this function at this point is greater than zero. By the condition $h > 0$, each point from $\mathbb{R}$ is covered by at most a finite number of basic functions.

It is easy to see that (substituting $s = \frac{t}{H}$)

$$\int_{-\infty}^{\infty} a_H(t)\, dt = \int_{-H}^{H} a_H(t)\, dt = \int_{-H}^{H} a\left(\frac{t}{H}\right) dt$$

$$= H \cdot \int_{-1}^{1} a(s)\, ds = H. \qquad (7.20)$$

If $h = H$, then an (h, H)-uniform fuzzy partition is called an h-uniform fuzzy partition.

Remark 1. The characterization "fuzzy" is used in the above definition for the sake of consistency with the original approach proposed in [136]. In particular, an (h, H)-uniform fuzzy partition simplifies proofs of many results in the theory of F-transforms.

Although basic functions a_k are actual membership functions of fuzzy sets, they are primary functions. In the sequel, we will be working with various other partitions (not necessarily fuzzy). All of them are unified by having a similar structure; i.e., all of them are translations of certain functions. A partition will be referred to as "fuzzy" only if it is a translation of a certain (rescaled) generating function; otherwise it is simply a "partition".

---

[6]For simplicity of representation, we assume that $t_0 = 0$.

The following lemma will be used in the sequel.

Lemma 4. Let $a : \mathbb{R} \to [0,1]$ be a generating function so that it is continuous, even, bell-shaped, vanishes outside $[-1,1]$ and fulfills $\int_{-1}^{1} a(t)\,dt = 1$. Then, the following is valid:

$$\frac{1}{2} \leq \|a\|^2 \leq 1, \tag{7.21}$$

where $\|a\|$ is the norm in $L_2([-1,1])$.

Proof. Obviously, $a \in L_2([-1,1])$ and for any $t \in [-1,1]$, $a^2(t) \leq a(t)$. Thus, we easily obtain $\|a\|^2 \leq 1$. The rest follows from the Cauchy-Schwarz inequality, so that

$$\left(\int_{-1}^{1} a(t)\,dt \right)^2 \leq 2 \int_{-1}^{1} a^2(t)\,dt.$$

□

In particular, if $a = a^{cos}$, then $\|a^{cos}\|^2 = \frac{3}{4}$.

7.10 DIRECT AND INVERSE F-TRANSFORM

In this section, we review formal notions of the direct and inverse F-transforms as introduced in [136] and extend the latter.

Assume that $x \in L_2(\mathbb{R})$ and $\{a_k, k \in \mathbb{Z}\}$ is an (h, H)-uniform fuzzy partition of $\mathbb{R}$, where $a_k(t) = a_H(t_k - t)$, a_H is the H-rescaled generating function a, and $t_k = k \cdot h$, $k \in \mathbb{Z}$, are nodes. The sequence $F[x] = \{X_k, k \in \mathbb{Z}\}$, where

$$X_k = \frac{\int_{-\infty}^{\infty} a_k(s) \cdot x(s)\,ds}{\int_{-\infty}^{\infty} a_k(s)\,ds}, \tag{7.22}$$

is the *(direct) F-transform* of x with respect to $\{a_k, k \in \mathbb{Z}\}$. Real numbers $X_k, k \in Z$, are the *F-transform components* of x. By the assumption of uniformity of the partition and by (7.20), the representation (7.22) of X_k can be rewritten and simplified as follows:

$$X_k = \frac{\int_{-\infty}^{\infty} a_H(t_k - s) \cdot x(s)\,ds}{\int_{-\infty}^{\infty} a_H(t_k - s)\,ds} = \frac{1}{H} \int_{-\infty}^{\infty} a_H(t_k - s) \cdot x(s)\,ds. \tag{7.23}$$

It is easy to see that if $x, y \in L_2(\mathbb{R})$, $\alpha \in \mathbb{R}$, then

$$F[x + y] = F[x] + F[y], \qquad (7.24)$$
$$F[\alpha x] = \alpha F[x].$$

The basic idea of the F-transform is to "capture" a local behavior[7] of an original function and characterize it by a certain value. It follows from (7.22) that the F-transform can be effectively computed for a rather wide class of functions. In particular, all continuous functions on compact domains can be originals of the F-transform.

Let $\mathbf{x} = (X_k, k \in \mathbb{Z})$ be an arbitrary sequence of reals and $\{a_k, k \in \mathbb{Z}\}$ be an (h, H)-uniform fuzzy partition of $\mathbb{R}$ with the H-rescaled generating function a. The following *inversion formula*

$$\hat{x}^F(t) = \frac{\sum_{k=-\infty}^{\infty} X_k \cdot a_k(t)}{\sum_{k=-\infty}^{\infty} a_k(t)}, \quad t \in \mathbb{R}, \qquad (7.25)$$

converts the sequence $\mathbf{x}$ into the real function $\hat{x}^F$ such that $\hat{x}^F :$ $\mathbb{R} \to \mathbb{R}$. Because the parameter h in an (h, H)-uniform fuzzy partition $\{a_k, k \in \mathbb{Z}\}$ of $\mathbb{R}$ is greater than zero, both sums in (7.25) contain only a finite number of non-zero summands. Because $H > \frac{h}{2}$, each point from $\mathbb{R}$ is covered by at least one basic function, so that the denominator in (7.25) is always non-zero. Therefore, the expression in (7.25) is correct.

We say that the function $\hat{x}^F$ is the *inverse F-transform of the sequence* $\mathbf{x} = (X_k, k \in \mathbb{Z})$ with respect to the fuzzy partition $\{a_k, k \in \mathbb{Z}\}$. If the sequence $\mathbf{x}$ consists of the F-transform components of some function x with respect to $\{a_k, k \in \mathbb{Z}\}$, then $\hat{x}^F$ is simply called the *inverse F-transform of* x.

It is easy to see that if $\mathbf{x}, \mathbf{y}$ are sequences of reals, $\alpha \in \mathbb{R}$, then

$$\widehat{(\mathbf{x} + \mathbf{y})}^F = \hat{x}^F + \hat{y}^F, \qquad (7.26)$$
$$\widehat{(\alpha \mathbf{x})}^F = \alpha \hat{x}^F.$$

The inverse F-transform $\hat{x}^F$ of a continuous function x can approximate x with an arbitrary precision. The desired quality of approximation can be achieved by a special choice of a partition. This fact can be easily proved using the technique introduced in [136].

---

[7]Component X_k of the F-transform of a function x is a weighted average of x in a vicinity of the node t_k covered by the basic function A_k.

7.11 RECONSTRUCTION FROM THE F-TRANSFORM COMPONENTS

The F-transform is the result of a linear correspondence between a set of functions from $L_2(\mathbb{R})$ and a set of sequences of reals. In general, the inversion formula does not define the inverse correspondence. In [136], it has been shown that the inverse F-transform can approximate a continuous function with an arbitrary precision. In practice, this means that for any level of precision, a fuzzy partition such that the quality of approximation by the corresponding inverse F-transform is lower than this level can be found. The important characteristic of this result is that it has been obtained for continuous functions with respect to the quality given by the maximal absolute difference between an original and approximating function.

In the later publications [132, 38], smooth approximations for functions from $L_2(\mathbb{R})$ by the inverse F-transforms were proposed. Moreover, it has been shown that the sequence of F-transform components of a function x can be transformed into another sequence of reals such that the inverse F-transform of the latter is the best approximation of x on the whole domain of this function. This result is different from that obtained in ([136]), where the best approximations of x by the F-transform components were proved on local subdomains covered by corresponding basic functions.

In [137], we showed even more; namely, the original function can be reconstructed from its F-transform components. Of course, this result was established for a narrower than $L_2(\mathbb{R})$ class of functions. In [137], our motivation stemmed from the Nyquist-Shannon-Kotelnikov reconstruction theorem and from the approach demonstrated in [38, 132].

7.12 FUZZY PARTITION CORRESPONDING TO THE COVERING OF $\mathbb{I}$

Let $N \in \mathbb{N}$ be sufficiently large and let the covering (7.11) of $\mathbb{I}$ by subintervals of ranks $k \in \{1, \ldots, 5\}$ described in Section 7.6 be fixed. In this section we construct a fuzzy partition of $\mathbb{I}$ such that for any rank k each subinterval of this rank corresponds to a 1-cut[8] of exactly one basic function.

---

[8]A 1-cut of a fuzzy set A is a subset A_1 of $\mathbb{R}$ such that $A_1 = \{x \in \mathbb{R} \mid A(x) = 1.\}$

Recall that the covering (7.11) of $\mathbb{I}$ consists of five sets $\mathcal{P}_1, \ldots, \mathcal{P}_5$ such that each set $\mathcal{P}_k, k = 1, \ldots, 5$, is a finite set of closed subintervals of rank k. For technical reasons, we are making slight changes to the definition of these sets. In what follows, the set $\mathcal{P}_k$ will consist of all subintervals of $\mathbb{I}$ remaining after removing all open intervals $(\frac{s}{N}, \frac{s+1}{N})$ from $0 \leq s < N$, $s \equiv k - 1 \,(\mathrm{mod}\ 5)$. The remaining closed intervals will be called *intervals of rank k*. For convenience, we denote intervals of rank k as $\mathbb{I}_{k,1}, \mathbb{I}_{k,2}, \ldots$, assuming that all elements of $\mathbb{I}_{k,i}$ are less than elements of $\mathbb{I}_{k,j}$ if $i < j$. It is worth noting that if $\mathbb{I}_{k,j} = [l_{k,j}, r_{k,j}]$, then it is possible that $l_{k,j} = r_{k,j}$. Lastly, the number of intervals of rank k, denoted N_k, is finite, but different for different k.

Lemma 5. Suppose $N \in \mathbb{N}$ and $\mathcal{P}_1, \ldots, \mathcal{P}_5$ are the sets of closed subintervals of rank k defined above. For each $k = 1, \ldots, 5$ we take $\mathcal{P}_k$ and define the corresponding set $\mathcal{A}_k$ of fuzzy sets $A_{k,j}$: $\mathbb{I} \to [0, 1]$, $j = 1, 2, \ldots, N_k$ using the following assignment:

- $A_{k,j} = 1$, if $x \in \mathbb{I}_{k,j}$;

- if $\mathbb{I}_{k,j} = [l_{k,j}, r_{k,j}]$ and $l_{k,j} \neq 0$, then on the interval $[l_{k,j} - \frac{1}{N}, l_{k,j}]$, $A_{k,j}(x)$ coincides with a linear function that goes through the points $(l_{k,j} - \frac{1}{N}, 0)$ and $(l_{k,j}, 1)$;

- if $\mathbb{I}_{k,j} = [l_{k,j}, r_{k,j}]$ and $r_{k,j} \neq 1$, then on the interval $[r_{k,j}, r_{k,j} + \frac{1}{N}]$, $A_{k,j}(x)$ coincides with a linear function that goes through the points $(r_{k,j}, 1)$ and $(r_{k,j} + \frac{1}{N}, 0)$;

- $A_{k,j} = 0$, if $x \in \mathbb{I} \setminus [\max(l_{k,j} - \frac{1}{N}, 0), \min(r_{k,j} + \frac{1}{N}, 1)]$.

Then

(i) for each $k = 1, \ldots, 5$, the set $\mathcal{A}_k = \{A_{k,j} : \mathbb{I} \to [0, 1], j = 1, 2, \ldots, N_k\}$ constitutes a fuzzy partition of $[0, 1]$ with the Ruspini condition;

(ii) the set $\mathcal{A} = \bigcup_{k=1}^{5} \mathcal{A}_k = \{A_{k,j} : \mathbb{I} \to [0, 1], k = 1, \ldots, 5, j = 1, 2, \ldots, N_k\}$ constitutes a "fuzzy 5-partition" of $[0, 1]$.

(iii) the following function $a : [-1, 1] \to \mathbb{R}$, where

$$a(t) = \begin{cases} 1, & \text{if } |t| \leq \frac{2}{3}, \\ 3(1 - |t|), & \text{otherwise,} \end{cases} \tag{7.27}$$

is generating for the fuzzy 5-partition $\mathcal{A}$, so that the elements of $\mathcal{A}$ coincide with h-translated and H-rescaled versions of the function a, where $h = \frac{1}{N}$ and $H = \frac{3}{N}$, i.e. they belong to the family of basic functions in $(\frac{1}{N}, \frac{3}{N})$-uniform a-generated fuzzy partition of $\mathbb{R}$.

(iv) the nodes $t_{k,j}$ of the fuzzy 5-partition $\mathcal{A}$ belong to the set $\{-1 + \frac{i}{N} \mid i = 1, \ldots, 2N + 2\}$.

7.13 INVERSE F-TRANSFORM OF A SEQUENCE WITH RESPECT TO THE FUZZY PARTITION $\mathcal{A}$

Let the assumptions of Section 1.12 be fulfilled. In accordance with Lemma 5 let $\mathcal{A}_k = \{A_{k,j} : \mathbb{I} \to [0,1], j = 1, 2, \ldots, N_k\}, k = 1, \ldots, 5$, constitute an a-generated fuzzy partition of $[0,1]$ with the Ruspini condition, where the generating function a is given by (7.27). Let moreover $\mathbb{Q}^{N_k}, k = 1, \ldots, 5$, be the set of sequences $\mathbf{v} = (v_j)$, $j = 1, \ldots, N_k$, where $\mathbb{Q}$ is the set of rationals in $[0,1]$. Then, the inversion formula (7.25) applied for $\mathbf{v}$, i.e.

$$\hat{v}(x) = \frac{\sum_{j=1}^{N_k} v_j A_{k,j}(x)}{\sum_{j=1}^{N_k} A_{k,j}(x)}, \tag{7.28}$$

converts the sequence $\mathbf{v}$ into a real continuous function $\hat{v} : [0,1] \to \mathbb{R}$, which is the inverse F-transform of $\mathbf{v}$ with respect to the fuzzy partition $\mathcal{A}_k$. Let $iFT(\mathbb{Q}^{N_k})$ denote the set of all inverse F-transforms of sequences $\mathbf{v} \in \mathbb{Q}^{N_k}$ with respect to the fuzzy partition $\mathcal{A}_k, k = 1, \ldots, 5$.

Theorem 3. Let $N \in \mathbb{N}$ and $\mathcal{A}_1, \ldots, \mathcal{A}_5$ be the a-generated fuzzy partitions of $[0,1]$ introduced in Lemma 5, where the generating function a is given by (7.27). For each $k = 1, \ldots, 5$, let $iFT(\mathbb{Q}^{N_k})$ be the corresponding set of all inverse F-transform of sequences $\mathbf{v} \in \mathbb{Q}^{N_k}$. Then,

(i) the inversion formula (7.28) for the functions from $iFT(\mathcal{A}_k)$ can be simplified to

$$\hat{v}(x) = \sum_{j=1}^{N_k} v_j A_{k,j}(x), \tag{7.29}$$

(ii) given $\mathbf{v} = (v_1, \dots v_{N_k}) \in \mathbb{Q}^{N_k}$, the corresponding function $\hat{v}$ represented in (7.29) acts as follows:

$$\hat{v}(x) = \begin{cases} v_j, & \text{if } x \in I_{k,j} = [l_{k,j}, r_{k,j}], \ j = 1, 2, \dots, N_k, \\ \ell(v_j, v_{j+1}), & \text{if } r_{k,j} < x < l_{k,j+1}, \ j = 1, 2, \dots, N_k - 1, \end{cases}$$

where $\ell(v_j, v_{j+1})$ is a linear function that connects points $(r_{k,j}, v_j)$ and $(l_{k,j+1}, v_{j+1})$, $j = 1, 2, \dots, N_k - 1$,

(iii) the set of all sets $iFT(\mathcal{A}_k)$ (where k is fixed and N varies) is dense in $C(\mathbb{I})$.

7.14 RELATIONSHIP TO THE KOLMOGOROV SUPERPOSITION THEOREM

In this section we reformulate Lemma 2 on the basis of Theorem 3. Then we will propose a new formulation of the Kolmogorov Superposition Theorem.

Lemma 6. Let us assume that the conditions of Lemma 2 are satisfied, i.e., some λ that satisfies Lemma 1 is fixed, function $f \in C(\mathbb{I}^2)$ such that $\|f\| = 1$ is chosen, and the set $U_f \subseteq [C(\mathbb{I})]^5$ is determined. We define a set $U_f^{iFT} \subseteq [C(\mathbb{I})]^5$ such that $(\phi_1, \dots, \phi_5) \in U_f^{iFT}$ if and only if

- there exists $N \in \mathbb{N}$ and a covering $\mathcal{P}_1, \dots, \mathcal{P}_5$ of the interval $\mathbb{I}$, as described in Section 1.12, such that every $\mathcal{P}_k$ contains N_k subintervals, $k = 1, \dots, 5$;

- there exists $(\phi_1^0, \dots, \phi_5^0) \in U_f$ such that for each $k = 1, \dots, 5$, ϕ_k coincides with the ϕ_k^0 on each interval of rank k in $\mathcal{P}_k$;

- there exist sequences $\mathbf{v}^k = (v_1^k, \dots v_{N_k}^k) \in \mathbb{Q}^{N_k}$ such that for each $k = 1, \dots, 5$, ϕ_k coincides with the inverse F-transform $\hat{v}^k$ of $\mathbf{v}^k$ with respect to fuzzy partition $\mathcal{A}_k$, corresponding to $\mathcal{P}_k$;

- there exists $g \in C(\mathbb{R})$, such that

$$|g(t)| \leq \frac{1}{7} \text{ for } t \in \mathbb{R},$$

and

$$\left| f(x, y) - \sum_{k=1}^{5} g(\phi_k(x) + \lambda \phi_k(y)) \right| < \frac{7}{8} \text{ for } (x, y) \in \mathbb{I}^2.$$

Then U_f^{iFT} is an open dense subset of $[C(\mathbb{I}^5)]$.

If we now replace Lemma 2 with Lemma 6, then preserving the sequence of all further steps given in Section 7.5 we obtain proof of the Theorem 2.

Remark 2. The new representation of the functions ϕ_k, $k = 1, \ldots, 5$ as the inverse F-transform $\hat{v}^k$ of the functions $\mathbf{v}^k$ with respect to the fuzzy partition $\mathcal{A}_k$, $k = 1, \ldots, 5$ uses the product operation.

7.15 GENERALIZED FUZZY PARTITIONS OF $\mathbb{I}^2$ AND THE INVERSE F-TRANSFORM REPRESENTATION OF FUNCTIONS $G \circ \Phi_K$

Let $N \in \mathbb{N}$, and let $\mathcal{A}_1, \ldots, \mathcal{A}_5$ be the fuzzy partitions of $[0, 1]$ described in Lemma 5. By this lemma, every set $\mathcal{A}_k = \{A_{k,j} : \mathbb{I} \to [0, 1], j = 1, 2, \ldots, N_k\}$, $k = 1, \ldots, 5$, consisting of elements of the a-generated fuzzy partitions of $[0, 1]$, where the generating function a is given by (7.27), satisfies the Ruspini condition.

For each $k = 1, \ldots, 5$, and a pair of indices (j, ℓ), where $j, \ell = 1, 2, \ldots, N_k$, we define a fuzzy set $A_{k,j,\ell}$ on $[0, 1]^2$ so that

$$A_{k,j,\ell}(x, y) = A_{k,j}(x) A_{k,\ell}(y).$$

Lemma 7. Let us assume that the conditions of Lemma 5 are satisfied and $\mathcal{A}_1, \ldots, \mathcal{A}_5$ are described there fuzzy partitions of $[0, 1]$. For each $k = 1, \ldots, 5$, we define a new set $\mathcal{A}_k^2$ of fuzzy sets defined on $[0, 1]^2$ such that $\mathcal{A}_k^2 = \{A_{k,j,\ell} : [0, 1]^2 \to [0, 1], j, \ell = 1, 2, \ldots, N_k\}$. We claim that $\mathcal{A}_k^2$ is a fuzzy partition of $[0, 1]^2$ satisfying the Ruspini condition, i.e., for all $x, y \in [0, 1]^2$,

$$\sum_{j=1}^{N_k} \sum_{\ell=1}^{N_k} A_{k,j,\ell}(x, y) = 1.$$

As in section 7.13, we introduce the concept of *inverse F-transform of a matrix with respect to the fuzzy partition of* $[0, 1]^2$. In particular, if $V = (v_{j,\ell})$ is a $(N_k \times N_k)$-matrix of reals, then the inverse F-transform of V with respect to $\mathcal{A}_k^2$ is a function $\hat{V}$ on $[0, 1]^2$ such that

$$\hat{V}(x, y) = \sum_{j=1}^{N_k} \sum_{\ell=1}^{N_k} v_{j,\ell} A_{k,j,\ell}(x, y).$$

In the above given representation, we made use of the fact that the fuzzy partition $\mathcal{A}_k^2$ satisfies the Ruspini condition.

Below, we will show that if the parameters of the representation formula (7.8) fulfill the assumptions of Lemma 8, then in the statement of Lemma 2, this formula can be rewritten as a sum of five inverse F-transforms of matrices with respect to the fuzzy partitions of $[0,1]^2$.

Lemma 8. Assume that some λ satisfying Lemma 1 is fixed, and a function $f \in C(\mathbb{I}^2)$ such that $\|f\| = 1$ is chosen. Let us assume that the conditions of Lemmas 2,6 are satisfied, so that the set $U_f \subseteq [C(\mathbb{I})]^5$, the function $g \in C(\mathbb{R})$, and the fuzzy partitions $\mathcal{A}_1, \ldots, \mathcal{A}_5$ of $[0,1]$ are determined. Let the functions Φ_k, $k = 1, \ldots, 5$, be defined as in (7.12). Then for each k, there exists a $(N_k \times N_k)$-matrix $V^k = (v_{j,\ell}^k)$ such that the inverse F-transform $\hat{V}^k$ of V^k with respect to fuzzy partition $\mathcal{A}_k^2$ of $[0,1]^2$ coincides with $g \circ \Phi_k$ on each rectangle $R_{k,j,\ell}$, $j = 1, \ldots$, of rank k, which is the Cartesian product of two intervals $\mathbb{I}_{k,j}$ and $\mathbb{I}_{k,\ell}$ of rank k.

Corollary 1. Let us assume that the conditions of Lemma 8 are satisfied. Then the inequality (7.10) can be rewritten as follows:

$$|f(x,y) - \sum_{k=1}^{5} \sum_{j=1}^{N_k} \sum_{\ell=1}^{N_k} v_{j,\ell}^k A_{k,j,\ell}(x,y)| < \frac{7}{8} \text{ for } (x,y) \in \mathbb{I}^2. \quad (7.30)$$

7.16 INVERSE F-TRANSFORM IN THE KOLMOGOROV SUPERPOSITION THEOREM

Having all the necessary technical details, we can formulate the main statement (for the case of $n = 2$ variables).

Theorem 4. For each continuous function $f : \mathbb{I}^2 \to \mathbb{R}$ and each $\varepsilon > 0$ there are five fuzzy partitions $\mathcal{A}_1, \ldots, \mathcal{A}_5$ of the interval $\mathbb{I}$ and five square real matrices $V^1, \ldots, V^5$ such that the sum of the inverse F-transforms $\hat{V}^k$ of the matrices V^k, with respect to the corresponding fuzzy partitions $\mathcal{A}_k^2$ approximates f with the accuracy of ε, i.e., for any $(x,y) \in \mathbb{I}^2$,

$$|f(x_1, x_2) - \sum_{k=1}^{5} \sum_{j=1}^{N_k} \sum_{\ell=1}^{N_k} v_{j,\ell}^k A_{k,j,\ell}(x,y)| < \varepsilon. \quad (7.31)$$

7.17 APPROXIMATION OF A CONTINUOUS FUNCTION BY ITS INVERSE F-TRANSFORM COMPUTED USING A "SHALLOW NEURAL NETWORK"

In this section we show how to use the inverse F-transform (7.25) representation and develop a "shallow" feedforward neural network with one hidden layer to compute (approximately) any continuous function given by a set of pairs consisting of arguments and their corresponding function values.

In Section 7.3, we showed how to design an approximating ANN based on the Cybenko approximation theorem. We proposed an initialization of the ANN parameters such that an ANN defined in this way not only computes function values given known inputs, but also predicts output values at intermediate points with good accuracy.

We now want to rely on the Kolmogorov superposition theorem, and in particular on its version using the inverse F-transform, and propose a different initialization of the ANN parameters than the above one. The most noticeable difference is the choice of the activation function type: sigmoid (Cybenko approximation) versus bell-shaped (inverse F-transform approximation). As a result of this choice, we will see that the number of hidden nodes in an ANN designed based on the inverse F-transform can be half as many. Thus, the ANN initialized based on the inverse F-transform theory outperforms the ANN initialized based on the Cybenko-type approximation.

In Figure 7.4 we show the result of approximate computation of the same function f given by (7.5) that we used in Section 7.3, where the proposed ANN had 200 hidden nodes. The initialization of the ANN parameters was performed based on the following set S of arguments-values belonging to the function f, defined by (7.5):

$$S = \left\{ \left(\frac{k\pi}{100}, e^{\frac{k\pi}{100}} \sin\left(\frac{k\pi}{10} \right) \right) \mid k = 1, \ldots, 100 \right\}.$$

This time, we designed an ANN with 100 hidden nodes using the inverse F-transform approximation scheme to initialize the ANN parameters.

We compare the approximation capabilities of two ANNs proposed in this section: one presented in Section 7.3 (sigmoid activation and 200 hidden nodes) and the latest one developed based

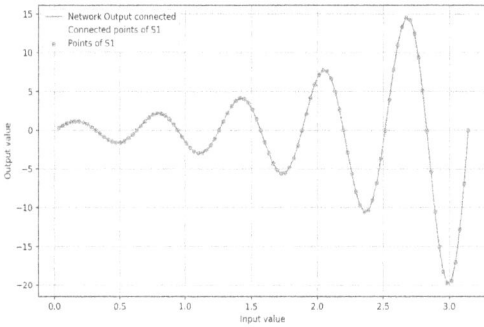

Figure 7.4 Plot of dataset S (dots) versus the values of function $f(x) = e^{\frac{\pi x}{100}} \sin\left(\frac{\pi x}{10}\right)$ computed at input points of dataset S using ANN initialized with inverse F-transform and $N = 100$ hidden neurons. Computed outputs are connected by a smooth line.

on the inverse F-transform (bell-shape activation and 100 hidden nodes). For comparison, we use both neural networks to predict a function f defined by (7.5) on both training and test sets, where the latter consists of points ("intermediate points") that lie between each pair of training points. By comparing (Table 7.1) the cumulative absolute differences of the predictions of the two neural networks proposed in this section on the training and test sets, we obtain comparative estimates of the quality of these neural networks.

In Figure 7.5 we show the visualization of the quality assessment of the results of two predictions computed at intermediate

Table 7.1 Accumulating Errors over training and testing sets

Approximation Model	Abs Error Train	Abs Error Prediction
inv FT	1.9806E-12	2.045E-12
Cybenko	33.081	69.228

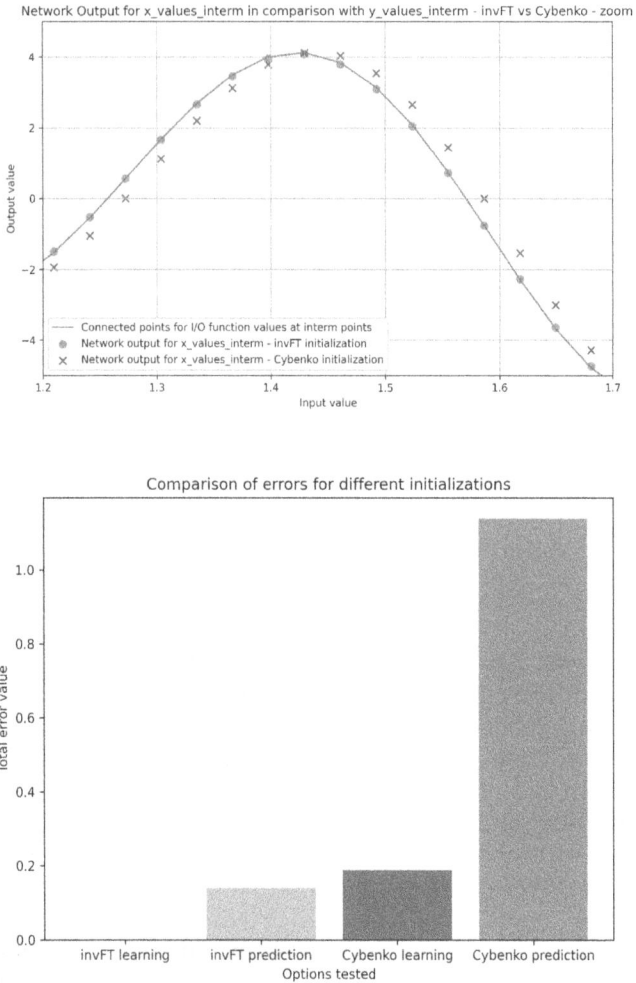

Figure 7.5 Upper: Plot of two predictions computed at intermediate points of S by two different ANNs trained on S but using different initializations associated with the Cybenko (crosses) and inverse F-transform (dots) approximations. Lower: Histogram of errors.

points of S by different ANNs trained on the same input-output data set S, but with two different initializations associated with the Cybenko and inverse F-transform approximation models. We see that the ANN initialized based on the inverse F-transform theory outperforms the ANN initialized based on the Cybenko-type approximation.

## CHAPTER 8

Fuzzy Modeling Methods in Processing of Long Time Series

Irina Perfilieva

University of Ostrava, Institute for Research and Applications of Fuzzy Modeling, 30. dubna 22, 701 03 Ostrava, Czech Republic

Vilém Novák

University of Ostrava, Institute for Research and Applications of Fuzzy Modeling, 30. dubna 22, 701 03 Ostrava, Czech Republic

Piotr Artiemjew

Faculty of Mathematics and Computer Science, University of Warmia and Mazury in Olsztyn, ul. Słoneczna 54, 10-710 Olsztyn, Poland

8.1 MOTIVATION

In this section, we will show that fuzzy modeling methods have the potential to compete with neural networks. We will focus on forecasting time series which is one of the tasks successfully solved by neural networks (see [91]) due to their ability to model

DOI: 10.1201/9781003515302-8

complex, nonlinear relationships in data. They can learn patterns occurring in the historical part of the time series directly and are particularly adept at handling the intricacies of time series data.

Neural networks can process vast amounts of data and learn from features that may not be immediately apparent. This is very useful in time series forecasting because the relevance of certain data points may only emerge after extensive analysis. Neural networks excel in identifying subtle, complex patterns and relationships that are often missed by traditional methods. Their problem, however, is the necessity of having large amounts of data at their disposal, high computational complexity, and the impossibility of explaining how the results were obtained.

In this section, we will demonstrate that many of the abilities of neural networks in processing the time series mentioned above also apply to methods based on fuzzy modeling techniques. We argue that the latter are sufficiently powerful when processing long time series, which otherwise is the realm of methods based on neural networks.

Unlike neural networks, our methods do not require much computational power. Moreover, they are transparent so that one can easily explain why and how the respective result was obtained. Such transparency stems from the theory of fuzzy natural logic mentioned further which includes a mathematical model of the meaning of selected linguistic expressions.

Let us also mention that forecasting using neural networks is usually bound to forecasting one future time moment. Of course, there is also the possibility to forecast times series over longer horizon[1] but it is computationally complex, and still, the horizon cannot be too long. Our methods, on the other hand, can forecast time series for a quite long future. We will demonstrate this below.

Fuzzy modeling methods can be also used in mining information from time series. Note that this problem is addressed by many authors (cf. [53, 105]) and is solved using both statistical as well as non-statistical methods.

We provide the mined information using expressions of natural language. Let us note that a similar task is solved also by neural networks. For example, in [151], the authors demonstrate how narration on the course of time series can be provided. The results of the generated narration look convincing. However, it relies on available text data when a neural network is learned and

---

[1]This is sometimes called *multi-horizon forecasting*.

so, unlike our methods, the narration is not based on the semantic model of the used expressions of natural language but on superficial similarity of texts.

In this section, we will briefly overview fuzzy modeling methods for processing of time series. The details can be found in several papers [115, 118, 121, 116, 114, 109, 120, 162] and in the book [119].

Recall that by a *fuzzy set* we understand a function $A : M \rightarrow [0,1]$ where M is a set. The value $A(m) \in [0,1]$ is a *truth value* of the proposition "$m \in M$ belongs to the fuzzy set A" and is called a *membership degree* of m in A.

8.2 FUZZY NATURAL LOGIC

Fuzzy natural logic (FNL) is a class of special formal theories of mathematical fuzzy logic whose goal is to model the reasoning of people based on using natural language. The main theory of FNL is that of *evaluative linguistic expressions* (see [112, 119]) that are expressions of natural language such as *small, medium, big, very short, more or less deep, quite roughly strong, extremely high*, etc.

In this section, we consider *simple evaluative expressions* having the form

$$\langle hedge \rangle TE\text{-adjective}.$$

The canonical TE-adjectives are *Ze* (zero), *Sm* (small), *Me* (medium) and Bi^2. The hedges that so far have a mathematical model are *Ex* (extremely), *Si* (significantly), *Ve* (very), *Ra* (rather), *ML* (more or less), *Ro* (roughly), *QR* (quite roughly), *VR* (very roughly) and a special hedge Δ (utmost). As a special case, we also consider the adjective *positive small* ^+Sm which excludes 0.

Another theory of FNL is that of *intermediate quantifiers*, which are expressions of natural language such as *all, most, almost all, many/much, several, a few, and some*. They are in FNL modeled using special logical formulas

$$(Q_{Ev}^{\forall} x)(B, A) \equiv (\exists z)[(\forall x)((B|z) x \Rightarrow Ax) \land Ev((\mu B)(B|z))], \tag{8.1}$$

$$(Q_{Ev}^{\exists} x)(B, A) \equiv (\exists z)[(\exists x)((B|z)x \land Ax) \land Ev((\mu B)(B|z))] \tag{8.2}$$

---

[2]It is clear that these TE-adjectives are only *canonical*. Their list is much larger and so far, it is not linguistically examined in detail.

where x is a variable for objects, z for fuzzy sets, A and B are formulas representing fuzzy sets[3] and Ev is an above-considered evaluative expression using which we evaluate the $size$ of the fuzzy set $B|z$ w.r.t. B. Note that Ev determines which quantifier is considered. In our theory, we consider the following quantifiers:

- "All B are A": $(Q^{\forall}_{Bi\,\Delta}x)(B, A)$,

- "Almost all B are A": $(Q^{\forall}_{Bi\,Ex}x)(B, A)$,

- "Most B are A": $(Q^{\forall}_{Bi\,Ve}x)(B, A)$,

- "Many/much B are A": $(Q^{\forall}_{\neg\,Sm}x)(B, A)$,

- "Not many/much B are A": $(Q^{\forall}_{\neg\,Sm}x)(B, A)$,

- "Several B are A": $(Q^{\forall}_{\neg(Sm\,Ve)}x)(B, A)$,

- "A few B are A": $(Q^{\forall}_{\neg(Sm\,Si)}x)(B, A)$,

- "Very few/little B are A": $(Q^{\forall}_{+Sm\,Ve}x)(B, A)$,

- "Some B are A": $(Q^{\exists}_{Bi\,\Delta}x)(B, A)$.

Either of the quantifiers (8.1) or (8.2) construes the sentence

$$\langle\text{Quantifier}\rangle\ B\ \text{are}\ A \tag{8.3}$$

where $\langle\text{Quantifier}\rangle$ is an intermediate quantifier in a linguistic form.

Let $B, Z, A \subseteq M$ be fuzzy sets and $\mu(B, B|Z)$ be a *measure of the size* of the Z-part of B[4].

Then the truth value of formula (8.1) is computed as follows:

$$(Q^{\forall}_{Ev}x)(B, A) =$$

$$\bigvee\left\{\bigwedge_{m\in M}((B|Z)(m) \to A(m)) \wedge Ev(\mu(B, B|Z))\ \middle|\ Z \subseteq M, B|Z \neq \varnothing\right\}. \tag{8.4}$$

---

[3]FNL is formulated in higher-order fuzzy logic, which makes it possible to deal with variables representing objects of various levels, i.e., pure objects x, fuzzy sets z, A, B or more complex functions. For more details, see [117, 113].

[4]The $B|Z$ is a fuzzy set consisting of all singletons from B that are equal to the corresponding singletons from Z.

In other words, it is the truth value of proposition (8.3). Note that we exclude the possibility that $B|Z = \varnothing$ since no quantifier considered above allows empty set as a universe of quantification.

Proposition 1. Let $B, A \subseteq M$ be nonempty sets. Then the truth value of the quantifier (8.4) is

$$(Q_{Ev}^{\vee} x)(B, A) = Ev(\mu(B, B|(B \cap A))). \tag{8.5}$$

Proof. Since A and B are sets, it only makes sense to consider cuts of B obtained using subsets $Z \subseteq M$. We may also avoid the case when $Z \cap B = \varnothing$ because then $B|Z = \varnothing$.

Let $Z \subseteq M \setminus A$. If $Z \cap B \neq \varnothing$ then $(B|Z)(m) \to A(m) = 0$ holds for all $m \in Z \cap B$ and so, $\bigwedge_{m \in M}((B|Z)(m) \to A(m)) = 0$.

Let $Z \cap A \neq \varnothing$ and $Z \cap (B \setminus A) \neq \varnothing$ and consider $m \in Z \cap B$ and $m \notin A$. Then $(B|Z)(m) \to A(m) = 0$ and so, $\bigwedge_{m \in M}((B|Z)(m) \to A(m)) = 0$.

Otherwise, let $Z \subseteq A$. If $B \cap A \subset Z \subseteq A$ then $B|Z = B \cap A$ and so, we may confine only to subsets of $B \cap A$. Let $Z \subseteq B \cap A$. Then for all $m \in M$ we have that $(B|Z)(m) \to A(m) = 1$, i.e., $\bigwedge_{m \in M}((B|Z)(m) \to A(m)) = 1$. Since μ is an increasing function and $\bigwedge_{m \in M}((B|(B \cap A))(m) \to A(m))$, $Z = B \cap A$ gives the greatest truth value in 8.4. From it follows (8.5). □

Note that for finite sets B, A, (8.5) can be obtained by

$$(Q_{Ev}^{\vee} x)(B, A) = Ev\left(\frac{|B \cap A|}{|B|}\right) \tag{8.6}$$

where $|\cdot|$ is the number of elements of the corresponding set.

8.3   TIME SERIES AND THEIR PROCESSING

A time series is a real or complex stochastic process that can be represented as follows:

$$X(t, \omega) = Tr(t) + C(t) + S(t) + R(t, \omega), \qquad t \in \mathbb{T}, \omega \in \Omega \tag{8.7}$$

where $\mathbb{T} = \{1, \ldots, p\}$ (in general, $\mathbb{T} \subset \mathbb{R}$) is a set of natural numbers interpreted as time moments, Ω is a set of elementary random events, Tr is a *trend* and $C(t)$ a *cycle* that are often joined into trend-cycle $TC(t) = Tr(t) + C(t)$. The $S(t)$ is a *seasonal* component that is a mixture of periodic functions and R is a random *noise* with the mean $\mathbf{E}(R(t, \omega)) = 0$ and variance $\mathbf{Var}(R(t, \omega)) < \sigma, t \in \mathbb{T}$. For a fixed $\omega \in \Omega$ the time series becomes a real (or complex) valued function X.

A powerful fuzzy modeling tool for processing of time series is the *fuzzy transform* described in Section 7. Given $h > 0$ and a fuzzy partition $\mathcal{A}_h = (A_1, \ldots, A_n)$, $n \geq 2$, defined over the set of nodes $c_0, \ldots, c_n \in \mathbb{T}^5$ such that $c_{k+1} = c_k + h$ where $h > 0$. First, we compute components $F_1[X], \ldots, F_n[X]$ and from them we compute inverse F-transform of X w.r.t. $\mathcal{A}_h$. The result is

$$\hat{X}_h(t) = \widehat{TC}_h(t) + \hat{S}_h(t) + \hat{R}_h(t), \qquad t \in \mathbb{T}. \qquad (8.8)$$

For mining information from time series, we consider components of degree 1, i.e.,

$$F_k^1[X](t) = \beta_k^0[X] + \beta_k^1[X](t - c_k), \qquad k = 0, \ldots, n. \qquad (8.9)$$

The coefficient $\beta_k^1[X]$ characterizes *slope* (first derivative) of X in the area characterized by the basic function $A_k \in \mathcal{A}_h$.

It can be proved that setting $h = dT$ for some d and a proper periodicity T occurring in X, the F-transform of the seasonal component $\hat{S}_h(t) \rightarrow 0$ and the noise $\hat{R}_h(t)$ is significantly reduced. Hence, the inverse F-transform (8.8) enables us to estimate the trend or trend-cycle:

$$TC_{h_{TC}} \approx \hat{X}_{h_{TC}} \quad \text{and} \quad Tr_{h_T} \approx \hat{X}_{h_T} \qquad (8.10)$$

where h_{TC} is set according to a periodicity T chosen from the middle of the list of found periodicities in the time series X, and similarly h_T from the longest ones.

Forecasting of time series is accomplished in two phases: first, we learn a linguistic description using which future components are estimated. The rules have the form

$$\text{IF } S_{i-1} \text{ is } \mathcal{A}_{i-1} \text{ AND } S_i \text{ is } \mathcal{A}_i \text{ THEN } S_{i+1} \text{ is } \mathcal{B}_{i+1} \qquad (8.11)$$

where S_i stands either for the components $F[X]_i$, their first differences $\Delta_i = F[X]_i - F[X]_{i-1}$, or the second ones $\Delta_i^2 = \Delta_i - \Delta_{i-1}$. Note the learned linguistic description also explains in natural language, how the forecast has been obtained. Let us remind that the subscript i refers to the *i-th component* of fuzzy partition. The future components $F[X]_{n+1}, \ldots$ are obtained from the learned linguistic description (8.11) using a special reasoning method called *Perception-based Logical Deduction* (PbLD). For the details see [119]. From them, using the inverse F-transform we obtain forecast of the trend or trend-cycle in (8.10).

---

[5]Usually but not necessarily, $c_0 = 1$ and $c_n = p$.

Forecast of the other components (cyclic or seasonal) can be computed, for example, using classical ARIMA, neural networks, or other methods.

8.4 INTERVALS OF MONOTONOUS BEHAVIOR AND SUMMARIZING INFORMATION FROM THEM

8.4.1 Evaluation and Quantification of Intervals of Monotonous Behavior

One of the possible problems in mining information from time series is identification of intervals with specific monotonous trend; for example, *stagnating, sharply increasing/decreasing, slightly increasing/decreasing*, etc. In [115], a simple procedure has been suggested whose result is decomposition of the time domain $\mathbb{T}$ into a set of adjacent intervals:

$$\mathcal{T} = \{\tau_i \mid i = 1, \ldots, s\}, \qquad \bigcup \mathcal{T} = \mathbb{T}$$

and consider a tangent $\beta^1[X|\tau]$ in an interval $\tau \in \mathcal{T}$ of the time series X computed using the F-transform. Each τ with the length $|\tau|$ is the *largest interval with the slope evaluated by one specific evaluative expression*.

A special position among methods for mining information from time series is held methods of automatic summarization (see, e.g., [29, 75, 105, 170]). We will apply the theory of intermediate quantifiers outlined above. The quantified summary of the information contained in one time series is the following:

In Q maximal intervals, trend of time series is $\mathcal{C}$

where $\mathcal{C}$ is an evaluative expression characterizing the slope $\beta^1[X|\tau]$. Though expressions $\mathcal{C}$ have a specific semantics, to simplify the task, we will translate them into evaluative expressions whose semantics is already developed. The way how it is done is in Table 8.1.

In our case, $B = \mathcal{T}$ and $A \subseteq \mathcal{T}$ is a set of maximal intervals τ whose trend is evaluated by $\mathcal{A}$ (for example, $\mathcal{A}$= "rapidly increasing"). Hence, we can apply Proposition 1 and compute the truth value of the quantifier Q above using (8.6).

Table 8.1 Slopes of time series characterized linguistically by $\mathcal{C}$ and their translation into known evaluative expressions. The expression *stagnating* is translated into Ze. The other expressions characterize in/de-creasing slope.

Expression $\mathcal{C}$	Translation	Expression $\mathcal{C}$	Translation
negligibly	$Sm\,Si$	fairly large	$Bi\,Ro$
very little	$Sm\,Ve$	quite large	$Bi\,Ra$
slightly	$Sm\,\bar{v}$	roughly	$Bi\,VR$
somewhat	$Sm\,Ra$	rapidly	$Bi\,Si$
a little	$Sm\,ML$	sharply	$Bi\,Ve$
clearly	$Me, Sm\,VR$	huge	$Bi\,Ex$
large	$Bi\,\bar{v}$		

8.4.2 Demonstration of Our Methods on Long Time Series

In this subsection, we will show that our methods can be applied also to long time series that are series with $|\mathbb{T}| \approx 1000$ or more. Processing of such series is more difficult. We will demonstrate our methods on three selected time series and for each of them we estimate their trend cycle, compute their forecast and show mining information including summarization from them.

Time series 1. This is time series with $|\mathbb{T}| = 2000$. Its condensed view is depicted in Figure 8.1. It contains estimation of the trend-cycle and its forecast. The detailed forecast is in Figure 8.2. The width of the basic functions forming the fuzzy partition was set to 71. The width of the testing part was set to 200. We can hardly imagine application of neural networks to forecast such a long horizon.

We also applied our algorithm to determination of maximal intervals of monotonous behavior. The total number of found intervals is 360. Their list is in Table 8.2 together with summarization using intermediate quantifiers that are accompanied by the respective truth values in the brackets. The linguistic evaluation of intervals is based on the context of maximal increase/decrease 100 over a time difference 10. From this we compute the context $[0, 4, 10]$ for the tangent $\beta^1[X|\tau]$.

One can see that it would be very difficult to identify these intervals manually. We can read the summarizing information from Table 8.1, for example, as follows:

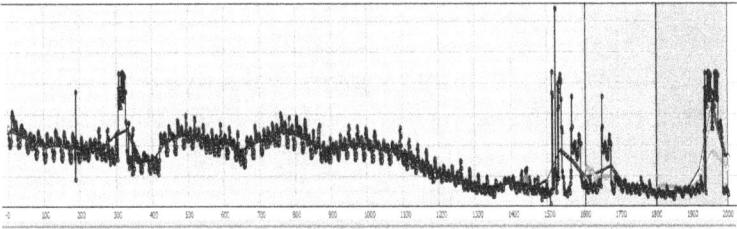

Figure 8.1 Condensed view on Time series 1, length $|\mathbb{T}| = 2000$. Notice that the estimated trend-cycle well fits the course of the time series.

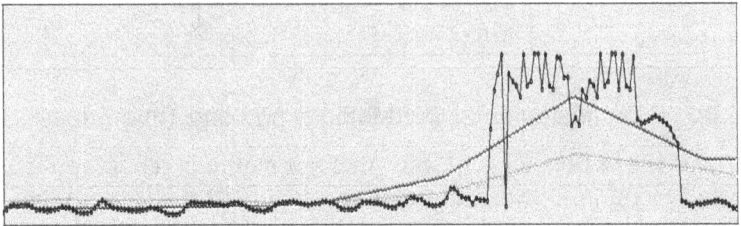

Figure 8.2 Time series 1, length $|\mathbb{T}| = 2000$. Detail of the forecast on testing part (equal to 200). The original time series and its computed trend-cycle are depicted by thin line.

> In Time series 1 there are *a few intervals* with *a little increasing trend.*

We chose here the quantifier with the highest truth degree.

The intervals in Table 8.2 seem to be too detailed and so, we may aggregate them. Namely, we will aggregate 1, 6, 10 into "a little increasing," 2, 7, 11 into "a little decreasing," 3, 8, 12 into "clearly increasing," 4, 9 into "clearly decreasing" and also "hugely in/decreasing" that did not occur in Table 8.2 because they alone did not contribute to the quantifiers. The result of such aggregation is in Table 8.3.

Time series 2. This is a time series with $|\mathbb{T}| = 10453$. Its condensed view is depicted in Figure 8.3. It contains an estimation of the trend-cycle and its forecast. The detailed forecast is depicted

Table 8.2 Maximal intervals of monotonous behavior detected in Time series 1 and quantification of their number using intermediate quantifiers. The corresponding truth values

No.	$\mathcal{A}$	Q
1.	**a little increasing:**	**A few (0.8)**, Several (0.45)
2.	a little decreasing:	A few (1), Several (0.85), Many (0.22)
3.	somewhat increasing:	A few (0.6), Several (0.25)
4.	somewhat decreasing:	A few (0.81), Several (0.5)
5.	stagnating:	Very few (0.98), A few (0.1)
6.	negligibly increasing:	Very few (0.9)
7.	negligibly decreasing:	Very few (0.94
8.	clearly increasing:	A few (1), Several (0.95), Many (0.45)
9.	clearly decreasing:	A few (1), Several (1), Many (0.55)
10.	very little increasing:	Very few (1)
11.	very little decreasing:	Very few (1)
12.	roughly decreasing:	Very few (1)

Table 8.3 Time series 1: Aggregated maximal intervals of monotonous behavior from Table 8.2.

No.	$\mathcal{A}$	Q
1.	stagnating:	Very few (0.98), A few (0.1)
2.	a little increasing:	A few (1), Several (0.95), Many (0.45)
3.	a little decreasing:	A few (1), Several (1), Many (0.7)
4.	clearly increasing:	A few (1), Several (1), Many (0.95)
5.	clearly decreasing:	A few (1), Several (1), Many (0.97)
6.	hugely increasing:	Very few (0.3)
7.	hugely decreasing:	Very few (0.2)

in Figure 8.4. The width of basic functions of the fuzzy partition was set to 650.

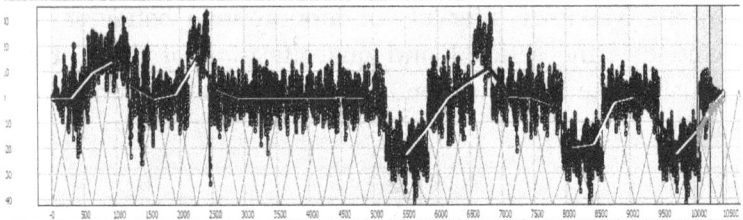

Figure 8.3 Condensed view of Time series 2, length $|\mathbb{T}|$ = 10453. One can see estimation of the trend-cycle together with the fuzzy partition (the width of basic functions is 650).

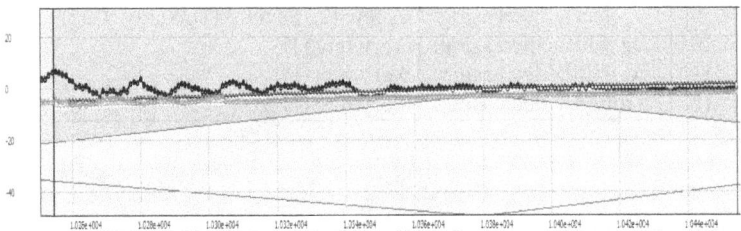

Figure 8.4 Time series 2, length $|\mathbb{T}|$ = 10453. Detail of the forecast on testing part (equal to 200) from Figure 8.3.

We also applied to this time series our algorithm for determination of maximal intervals of monotonous behavior. The total number of found intervals is 1296. Among them, 231 intervals are stagnating, 532 increasing and 533 decreasing. The linguistic evaluation of intervals is based on the context of maximal in/decrease 100 over time difference 8. From this we compute context $[0, 5, 12.5]$ for the tangent $\beta^1[X|\tau]$. Since the number of intervals is large, similarly as above, we decided to aggregate too detailed intervals into slightly larger ones. Their list is in Table 8.4 together with summarization using intermediate quantifiers and the corresponding truth values in brackets.

Our algorithm for finding intervals of monotonous behavior was also applied to the estimated trend-cycle. The list of found intervals is in Table 8.5.

Table 8.4 Time series 2: Aggregated maximal intervals of monotonous behavior .

No.	$\mathcal{A}$	Q
1.	stagnating:	A few (1), Several (1), Many (0.55)
2.	a little increasing:	A few (1), Several (1), Many (1)
3.	a little decreasing:	A few (1), Several (1), Many (1)
4.	clearly increasing:	A few (1), Several (1), Many (1)
5.	clearly decreasing:	A few (1), Several (1), Many (1)
6.	rapidly increasing:	Very few (0.97)
7.	rapidly decreasing:	Very few (1)
8.	hugely increasing:	A few (0.55), Several (0.9), Many (0.28)
9.	hugely decreasing:	A few (1), Several (0.93), Many (0.35)

Table 8.5 Time series 2: List of detected intervals of monotonous behavior in the estimated trend-cycle.

Interval	Evaluation
$[9713, 10454]$	clearly increasing
$[9066, 9713]$	fairly large decrease
$[8137, 9066]$	a little increasing
$[7396, 8137]$	roughly decreasing
$[7125, 7396]$	stagnating
$[6760, 7125]$	roughly decreasing
$[5549, 6760]$	**clearly increasing**
$[5466, 5549]$	very little increasing
$[4819, 5466]$	roughly decreasing
$[2950, 4819]$	stagnating
$[2303, 2950]$	somewhat decreasing
$[2220, 2303]$	clearly decreasing
$[1667, 2220]$	fairly large increase
$[1584, 1667]$	somewhat increasing
$[937, 1584]$	clearly decreasing
$[290, 937]$	clearly increasing
$[1, 290]$	stagnating

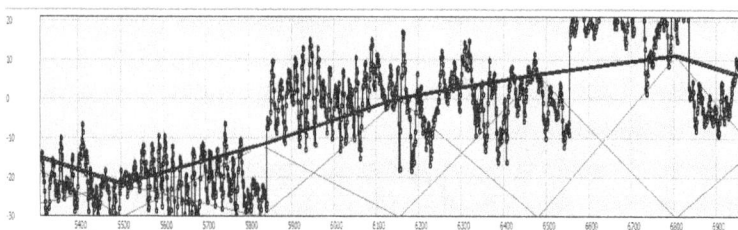

Figure 8.5 Time series 2, length $|\mathbb{T}| = 10453$. Example of detected increasing interval $[5549, 6760]$ in the estimated trend-cycle.

The bold-face font is used to emphasize an example of increasing interval $[5549, 6760]$ in the estimated trend-cycle from Figure 8.3. It is in detail depicted in Figure 8.5.

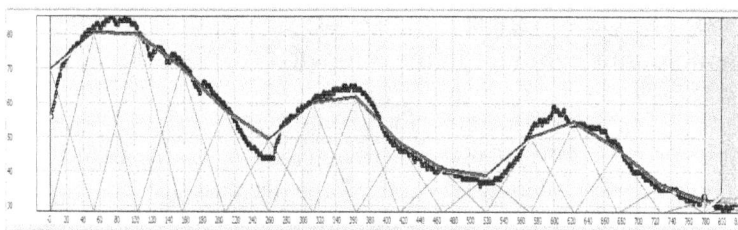

Figure 8.6 Time series 3, length $|\mathbb{T}| = 820$, together with estimation of the trend-cycle that was computed using F-transform based on the depicted fuzzy partition. On the right of the figure are validation and testing sets and the forecast.

Time series 3. This is a slightly shorter time series with $|\mathbb{T}| = 820$. Its condensed view is in Figure 8.6. One can see estimation of the trend-cycle and its forecast with forecasting horizon 20. The detail of the latter is in Figure 8.7.

The forecast was obtained using the linguistic description in Table 8.6 that was learned from the data. Recall that the subscripts refer to components of the fuzzy partition. The $\Delta_{i+1} = F[X]_{i+1} - F[X]_i$ is the first (future) difference and $\Delta_i^2 = \Delta_i - \Delta_{i-1}$,

Table 8.6 Time series 3: linguistic description using which forecast of the trend cycle was estimated.

No.	Δ_i^2 & $\Delta_{i-1}^2 \Rightarrow \Delta_{i+1}$
1	-ra me &-ml me ⇒ -ra bi
2	-ro sm & -ra me ⇒ -vr bi
3	qr sm & -ro sm ⇒ ro bi
4	ex bi & qr sm ⇒ ml sm
5	-ra me & ex bi ⇒ -si bi
6	-ro bi & -ra me ⇒-ml me
7	ml me & -ro bi ⇒-ml sm
8	ra me & ml me ⇒ ml bi
9	vr bi & ra me ⇒ ra me
10	-ra me & vr bi ⇒-ml me
11	-vr bi & -ra me ⇒ -ro bi
12	-ml sm & -vr bi ⇒-ml me

$\Delta_{i-1}^2 = \Delta_{i-1} - \Delta_{i-2}$ are second differences of the components of the applied F-transform. The future difference Δ_{i+1} was estimated using the PbLD inference on the basis of the linguistic description from Table 8.6.

This description gives us information about the way how the forecast was obtained. It also enables us to test how the forecast would be changed if we apply different input data. This is demonstrated in Figure 8.9 which compares the forecast obtained on the basis of given time series with two forecasts if the time series were slightly modified.

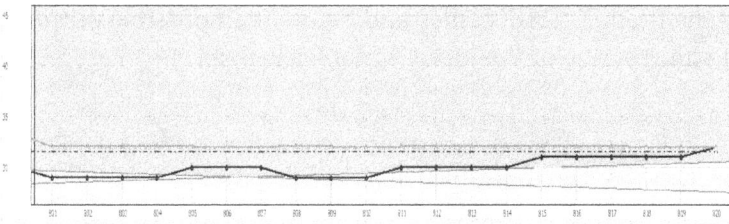

Figure 8.7 Time series 3, length $|\mathbb{T}| = 820$. Detail of the forecast in the testing set.

(a)

(b)

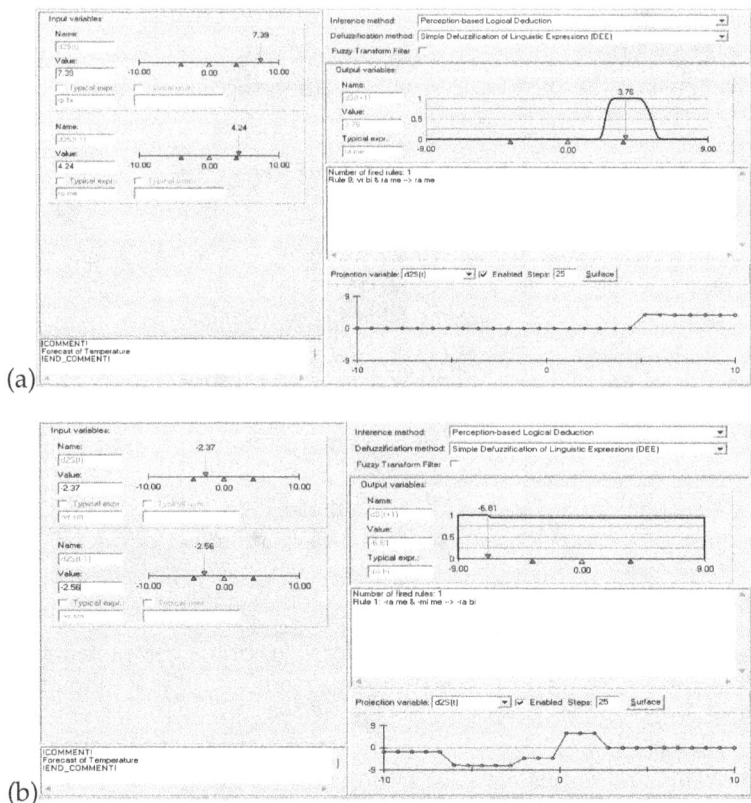

Figure 8.8 Time series 3: PbLD Inference giving (a) higher and (b) lower forecast. On the left are two sliders for entering the input values. On the right is an output fuzzy sets, marked output value obtained using defuzzification. Below is the fired IF-THEN rules and below it all possible outputs if value of one of the input variables is fixed.

In Figure 8.8 we demonstrate two inferences using PbLD deduction based on the learned linguistic description from Table 8.6. The left part contains two sliders for entering the input values. In the right part, the resulting output fuzzy set is depicted with marked output value after defuzzification. Below is a fired rule and finally all possible outputs if value of one of the input

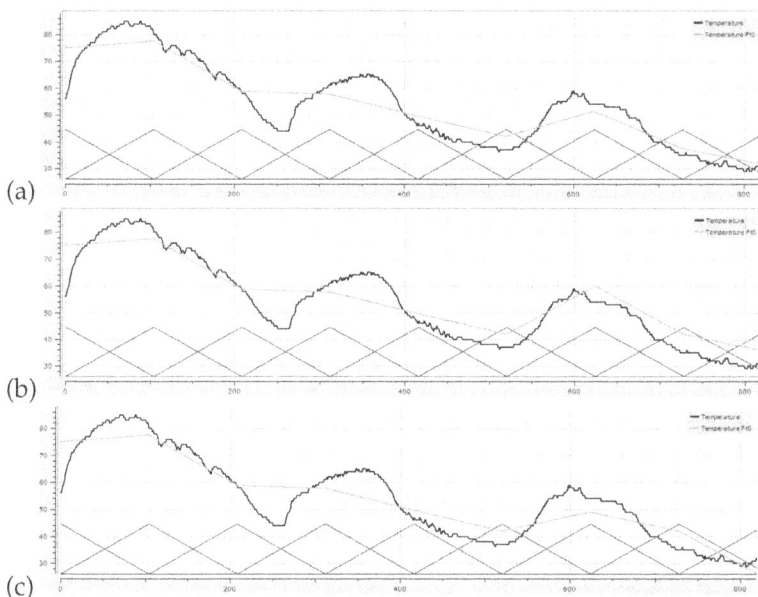

(a)

(b)

(c)

Figure 8.9   Comparison of possible forecasts of Time series 3: (a) original from the data, (b) higher, (c) lower. The forecasts (b), (c) are based on the linguistic description from Table 8.6 after entering inputs from Figure 8.8.

variables is fixed. The result how the forecasts are changed is in Figure 8.9.

8.4.3   Other sophisticated summarizing information on time series

We can generate summarizing information either on one time series or on a set of them (this may contain hundreds of time series). In the following examples, we demonstrate possible sentences that can be generated using methods of fuzzy natural logic.

- In *most (almost all, many, a few)* time intervals, *increasing (decreasing)* trend is followed by *rapidly (very slightly, sharply, clearly) decreasing (increasing)* one.

- *Short* periods of *sharp increase* of trend are *often* followed by *long periods* of *stagnation*.

- There are many time series with long *stagnating* (increasing, very little decreasing) trend.

- There is period of a *huge* (*slight, clear*) decrease in trend of *many* time series.

- *Most* (*many, several*) analyzed time series *stagnated* recently but their expected trend is *slightly increasing*.

- Trend of *most* of the inspected time series in the past 3 months was *sharply increasing*.

- *Most* time series are expected to *increase* in the following 6 months.

- During the last year, *several* time series were *sharply increasing* more than once.

8.5   CONCLUSION

In this chapter, we focused on two areas: neural networks and fuzzy modeling methods. In Section 7, we analyzed representations of continuous functions as realizations of the input-output relation, so that the computations corresponding to them can be performed by suitable ANNs. In this section, we have shown how a particular representation method facilitates popular computations using neural networks.

We started with presentation of the classical ideas due to Kolmogorov and Cybenko and then we have shown how they contribute to the popular computation with neural networks. Our results also contribute to better understanding of the meaning of the parameters involved in it. Then, we introduced a new type of universal approximation of continuous functions that follows from the classical formulas and extends them based on new principles that follow from partitioning the underlying domain into a set of fuzzy subsets. As a result, the ANN initialized on the basis of this new type of universal approximation (called the inverse F-transform) outperforms the ANN initialized on the basis of the Cybenko-type approximation.

In Section 8, we have shown that processing of long time series using neural networks can be effectively realized using methods of fuzzy modeling which are much less computationally demanding and provide a good estimation of the trend cycle and its forecast. Moreover, the latter is accompanied by a linguistic description explaining how was the forecast obtained. Our methods can also be used for mining information from time series in a form of automatically generated expressions of natural language. We demonstrated the results of the mentioned methods on three examples of long time series.

ACKNOWLEDGMENT

The work on the chapters has been supported by the Polish National Agency for Academic Exchange Strategic Partnership Programme under Grant No. BPI/PST/2021/1/00031.

The work of Irina Perfilieva was partially supported by the project AIMet4AI No. CZ.02.1.01/0.0/0.0/17-049/0008414.

The work of Vilém Novák has been partially produced with the financial support of the European Union under the REFRESH "Research Excellence For REgion Sustainability and High-tech Industries" project number CZ.10.03.01/00/22-003/0000048 via the Operational Programme Just Transition.

## CHAPTER 9

Conclusions

Agnieszka Niemczynowicz and Radosław Kycia

Faculty of Computer Science and Telecommunications, Cracow University of Technology, Warszawska 24, 31-155 Kraków, Poland

The theoretical foundations of Artificial Neural Networks (ANNs) were established in the 20th century, beginning with the McCulloch–Pitts perceptron, progressing through the development of the multilayer perceptron, and culminating in modern architectures where the backpropagation algorithm proved effective for training. However, this technology has only recently experienced a surge in popularity, driven by the advent of sufficiently powerful Graphics Processing Units (GPUs) capable of training NNs with vast numbers of trainable parameters. Large Language Models (LLMs) are now capable of assisting in numerous human activities, and the demand for other AI-powered systems continues to grow. Data analysis and neural network-based solutions now perform as well as — or often better than — classical algorithms in nearly every domain. As a result, the early 21st century can be considered the rise of the *data-driven algorithms* paradigm.

The core idea behind neural networks is remarkably simple: construct a complex (cascaded[1]) system built from small, simplified computational elements. It turns out that this approach is highly flexible and can be successfully applied to virtually any type of data. In the most basic theoretical sense, such complex systems serve as universal approximators for specific classes of functions.

---

[1]For feedforward neural networks.

DOI: 10.1201/9781003515302-9

The current excitement around neural networks continues, and the discipline is evolving in so many directions that it's increasingly difficult to track. Much of this development is fueled by the demands of major tech companies seeking innovations that can be implemented in commercial products. So far, progress has largely kept pace with these high expectations. However, the history of AI reminds us that unmet expectations — despite heavy investment — can lead to a paradigm shift. This was the case during the two so-called "AI winters" when research on neural networks was largely abandoned in favor of alternative approaches. Hopefully, this fate can be avoided in the current wave of AI advancement.

This volume presents selected modern perspectives on the ongoing development of neural networks. We begin with the basics of classical neural networks and explore their applications, including their integration into Large Language Models and their optimization using genetic algorithms.

We then move on to the topic of complex- and quaternion-valued neural networks, presenting their foundations and a variety of real-world applications. A common guiding principle in these architectures is to align the natural dimensionality of the data with the dimensionality of the algebra. One example is DNA encoding, where the four chemical bases — A, C, T, G — can be represented as the four components of a quaternion. Another example involves encoding images, where the Red, Green, and Blue (RGB) color channels — along with an artificial Alpha channel — can be mapped into a four-dimensional algebra such as quaternions[2].

Finally, we delve into theoretical aspects of the universal approximation theorem and explore its fascinating connection with the fuzzy transform, which itself serves as a powerful approximation tool. We continue the fuzzy path by examining how fuzzy natural logic can be employed to decode the behavior of time series data.

---

[2]We also added an artificial Alpha channel to form an ARGB representation with four components.

At some point, we must conclude — even if the topic is far from exhausted. Nevertheless, we hope the contents of this volume will bring much joy to the curious reader in exploring the fascinating world of hypercomplex neural networks and their related directions. We believe that this book contains enough depth and insight to inspire readers to embark on their own research journey: *"A journey of a thousand miles begins with a single step."*[3].

---

[3]Laozi

Bibliography

[1] Martín Abadi et al. *TensorFlow: Large-Scale Machine Learning on Heterogeneous Systems*. Software available from tensorflow.org. 2015. URL: https://www.tensorflow.org/.

[2] Behzad Abbasi, Vahid Majidnezhad, and Seyedali Mirjalili. "ADE: advanced differential evolution". In: *Neural Computing and Applications* (2024). ISSN: 1433-3058. DOI: 10.1007/s00521-024-09669-z. URL: http://dx.doi.org/10.1007/s00521-024-09669-z.

[3] Debasis Acharya and Dushmanta Kumar Das. "A novel Human Conception Optimizer for solving optimization problems". In: *Scientific Reports* 12.1 (2022). ISSN: 2045-2322. DOI: 10.1038/s41598-022-25031-6. URL: http://dx.doi.org/10.1038/s41598-022-25031-6.

[4] Josh Achiam et al. "Gpt-4 technical report". In: *arXiv preprint arXiv:2303.08774* (2023).

[5] Jeffrey O. Agushaka et al. "Greater cane rat algorithm (GCRA): A nature-inspired metaheuristic for optimization problems". In: *Heliyon* 10.11 (2024), e31629. ISSN: 2405-8440. DOI: 10.1016/j.heliyon.2024.e31629. URL: http://dx.doi.org/10.1016/j.heliyon.2024.e31629.

[6] Tom B. Brown et al. *Language Models are Few-Shot Learners*. 2020. arXiv: 2005.14165 [cs.CL].

[7] Alessia Amelio and Clara Pizzuti. "A Genetic Algorithm for Color Image Segmentation". In: *Lecture Notes in Computer Science*. Springer Berlin Heidelberg, 2013, pp. 314–323. ISBN: 9783642371929. DOI: 10.1007/978-3-642-37192-9_32. URL: http://dx.doi.org/10.1007/978-3-642-37192-9_32.

[8] Anthropic. *Introducing Claude*. 2023. URL: https://www.anthropic.com/index/introducing-claude.

[9] "Neural Networks in Complex Algebra". In: *Neural Networks in Multidimensional Domains*. Ed. by P. Arena et al. Vol. 234. Lecture Notes in Control and Information Sciences. London: Springer, 1998. DOI: 10.1007/BFb0047685. URL: https://doi.org/10.1007/BFb0047685.

[10] Paolo Arena et al. *Neural networks in multidimensional domains: fundamentals and new trends in modelling and control.* Springer, 1998.

[11] Paolo Arena et al. "On the capability of neural networks with complex neurons in complex valued functions approximation". In: *1993 IEEE international symposium on circuits and systems*. IEEE. 1993, pp. 2168–2171.

[12] V.I. Arnold. "On functions of three variables (in Russian)". In: *Dokl. Akad. Nauk SSSR* 114 (1957), pp. 679–681.

[13] Esmaeil Atashpaz-Gargari and Caro Lucas. "Imperialist competitive algorithm: An algorithm for optimization inspired by imperialistic competition". In: *2007 IEEE Congress on Evolutionary Computation*. IEEE, 2007. DOI: 10.1109/cec.2007.4425083. URL: http://dx.doi.org/10.1109/CEC.2007.4425083.

[14] Enes Ayan. "Genetic Algorithm-Based Hyperparameter Optimization for Convolutional Neural Networks in the Classification of Crop Pests". In: *Arabian Journal for Science and Engineering* 49.3 (2023), pp. 3079–3093. ISSN: 2191-4281. DOI: 10.1007/s13369-023-07916-4. URL: http://dx.doi.org/10.1007/s13369-023-07916-4.

[15] Tummala. S. L. V. Ayyarao et al. "War Strategy Optimization Algorithm: A New Effective Metaheuristic Algorithm for Global Optimization". In: *IEEE Access* 10 (2022), pp. 25073–25105. ISSN: 2169-3536. DOI: 10.1109/access.2022.3153493. URL: http://dx.doi.org/10.1109/ACCESS.2022.3153493.

[16] Thomas Back and Frank Hoffmeister. "Basic aspects of evolution strategies". In: *Statistics and Computing* 4.2 (1994). ISSN: 1573-1375. DOI: 10.1007/bf00175353. URL: http://dx.doi.org/10.1007/BF00175353.

[17] Luca Baronti and Marco Castellani. *A Python Benchmark Functions Framework for Numerical Optimisation Problems.* 2024. arXiv: 2406.16195.

[18] Emily M Bender et al. "On the Dangers of Stochastic Parrots: Can Language Models Be Too Big?" In: *Proceedings of the 2021 ACM Conference on Fairness, Accountability, and Transparency.* 2021, pp. 610–623.

[19] Gerardo Beni and Jing Wang. "Swarm Intelligence in Cellular Robotic Systems". In: *Robots and Biological Systems: Towards a New Bionics?* Springer Berlin Heidelberg, 1993, pp. 703–712. ISBN: 9783642580697. DOI: 10.1007/978-3-642-58069-7_38. URL: http://dx.doi.org/10.1007/978-3-642-58069-7_38.

[20] Ms. Trupti Bhoskar et al. "Genetic Algorithm and its Applications to Mechanical Engineering: A Review". In: *Materials Today: Proceedings* 2.4–5 (2015), pp. 2624–2630. ISSN: 2214-7853. DOI: 10.1016/j.matpr.2015.07.219. URL: http://dx.doi.org/10.1016/j.matpr.2015.07.219.

[21] Julian Blank and Kalyanmoy Deb. "Pymoo: Multi-Objective Optimization in Python". In: *IEEE Access* 8 (2020), pp. 89497–89509. ISSN: 2169-3536. DOI: 10.1109/access.2020.2990567. URL: http://dx.doi.org/10.1109/ACCESS.2020.2990567.

[22] Rishi Bommasani et al. "On the Opportunities and Risks of Foundation Models". In: *arXiv preprint arXiv:2108.07258* (2021).

[23] N. Boullé et al. "Classification of chaotic time series with deep learning". In: *Physica D: Nonlinear Phenomena* 403 (2020), p. 132261. DOI: 10.1016/j.physd.2019.132261. URL: https://doi.org/10.1016/j.physd.2019.132261.

[24] James Bradbury et al. *JAX: composable transformations of Python+NumPy programs.* Version 0.3.13. 2018. URL: http://github.com/jax-ml/jax.

[25] Michael M Bronstein et al. "Geometric deep learning: Grids, groups, graphs, geodesics, and gauges". In: *arXiv preprint arXiv:2104.13478* (2021).

[26] Robert Brown. "On generalized Cayley-Dickson algebras". In: *Pacific Journal of Mathematics* 20.3 (1967), pp. 415–422.

[27] Tom B. Brown et al. "Language Models are Few-Shot Learners". In: *Advances in Neural Information Processing Systems* 33 (2020), pp. 1877–1901.

[28] Hieronymus Cardano. *Artis magnae, sive de regulis algebraicis, liber unus.* Joh. Petreius, 2011.

[29] R. Castillo-Ortega, N. Marín, and D. Sánchez. "A Fuzzy Approach to the Linguistic Summarization of Time Series". In: *Multiple-Valued Logic and Soft Computing* 17.2-3 (2011), pp. 157–182.

[30] C.H.D.C. de Castro and F.A.L. Aiube. "Forecasting inflation time series using score-driven dynamic models and combination methods: The case of Brazil". In: *Journal of Forecasting* 42.2 (2023), pp. 369–401. DOI: 10.1002/for.2908. URL: https://doi.org/10.1002/for.2908.

[31] Razvan Cazacu. "Comparative Study between the Improved Implementation of 3 Classic Mutation Operators for Genetic Algorithms". In: *Procedia Engineering* 181 (2017), pp. 634–640. ISSN: 1877-7058. DOI: 10.1016/j.proeng.2017.02.444. URL: http://dx.doi.org/10.1016/j.proeng.2017.02.444.

[32] Eunice YS Chan and Robert M Corless. "Chaos game representation". In: *SIAM Review* 65.1 (2023), pp. 261–290.

[33] Chiranjit Changdar, G.S. Mahapatra, and Rajat Kumar Pal. "An improved genetic algorithm based approach to solve constrained knapsack problem in fuzzy environment". In: *Expert Systems with Applications* 42.4 (2015), pp. 2276–2286. ISSN: 0957-4174. DOI: 10.1016/j.eswa.2014.09.006. URL: http://dx.doi.org/10.1016/j.eswa.2014.09.006.

[34] François Chollet et al. *Keras.* https://keras.io. 2015.

[35] Bastien Chopard and Marco Tomassini. *An Introduction to Metaheuristics for Optimization.* Springer International Publishing, 2018. ISBN: 9783319930732. DOI: 10.1007/978-3-319-93073-2. URL: http://dx.doi.org/10.1007/978-3-319-93073-2.

[36] Matthieu Courbariaux et al. *Binarized Neural Networks: Training Deep Neural Networks with Weights and Activations Constrained to +1 or -1.* 2016. DOI: 10.48550/ARXIV.1602.02830. URL: https://arxiv.org/abs/1602.02830.

[37] J. Vaca-Rubio Cristian et al. "Kolmogorov-Arnold Networks (KANs) for Time Series Analysis". In: *arXiv: 2405.08790* (2024).

[38] J.F. Crouzet. "Fuzzy projection versus inverse fuzzy transform as sampling/interpolation schemes". In: *Fuzzy Sets and Systems* 193 (2011), pp. 108–121.

[39] Yarens J. Cruz et al. "Ensemble of convolutional neural networks based on an evolutionary algorithm applied to an industrial welding process". In: *Computers in Industry* 133 (2021), p. 103530. ISSN: 0166-3615. DOI: 10.1016/j.compind.2021.103530. URL: http://dx.doi.org/10.1016/j.compind.2021.103530.

[40] G. Cybenko. "Approximation by superpositions of a sigmoidal function". In: *Math. Control Signals Systems* 2/4 (1989), pp. 303–314.

[41] George Cybenko. "Approximation by superpositions of a sigmoidal function". In: *Mathematics of control, signals and systems* 2.4 (1989), pp. 303–314.

[42] Arman Daliri et al. "The water optimization algorithm: a novel metaheuristic for solving optimization problems". In: *Applied Intelligence* 52.15 (2022), pp. 17990–18029. ISSN: 1573-7497. DOI: 10.1007/s10489-022-03397-4. URL: http://dx.doi.org/10.1007/s10489-022-03397-4.

[43] Bikash Das, V. Mukherjee, and Debapriya Das. "Student psychology based optimization algorithm: A new population based optimization algorithm for solving optimization problems". In: *Advances in Engineering Software* 146 (2020), p. 102804. ISSN: 0965-9978. DOI: 10.1016/j.advengsoft.2020.102804. URL: http://dx.doi.org/10.1016/j.advengsoft.2020.102804.

[44] Khatereh Davoudi and Parimala Thulasiraman. "Evolving convolutional neural network parameters through the genetic algorithm for the breast cancer classification problem". In: *SIMULATION* (2021), p. 003754972199603. ISSN: 1741-3133. DOI: 10.1177/0037549721996031. URL: http://dx.doi.org/10.1177/0037549721996031.

[45] Kusum Deep and Manoj Thakur. "A new crossover operator for real coded genetic algorithms". In: *Applied Mathematics and Computation* 188.1 (2007), pp. 895–911. ISSN: 0096-3003. DOI: 10.1016/j.amc.2006.10.047. URL: http://dx.doi.org/10.1016/j.amc.2006.10.047.

[46] Patrick J Deschavanne et al. "Genomic signature: characterization and classification of species assessed by chaos game representation of sequences." In: *Molecular biology and evolution* 16.10 (1999), pp. 1391–1399.

[47] Jacob Devlin et al. "BERT: Pre-training of Deep Bidirectional Transformers for Language Understanding". In: *Preceedings of the 2019 Conference of the North AMerican Chapter of the Association for Computational Linguistics: Human Language Technologies, Volume 1* (2019).

[48] Shifei Ding, Chunyang Su, and Junzhao Yu. "An optimizing BP neural network algorithm based on genetic algorithm". In: *Artificial Intelligence Review* 36.2 (2011), pp. 153–162. ISSN: 1573-7462. DOI: 10.1007/s10462-011-9208-z. URL: http://dx.doi.org/10.1007/s10462-011-9208-z.

[49] Editorial. "Embedding AI in biology". In: *Nat Methods 21* (2024). DOI: 10.1038/s41592-024-02391-7.

[50] Michael Eisenstein. "Foundation models build on ChatGPT tech to learn the fundamental language of biology". In: *Nature biotechnology* 42.9 (2024), pp. 1323–1325.

[51] Fatima Zahrae El-Hassani et al. "A New Optimization Model for MLP Hyperparameter Tuning: Modeling and Resolution by Real-Coded Genetic Algorithm". In: *Neural Processing Letters* 56.2 (2024). ISSN: 1573-773X. DOI: 10.1007/s11063-024-11578-0. URL: http://dx.doi.org/10.1007/s11063-024-11578-0.

[52] Félix-Antoine Fortin et al. "DEAP: Evolutionary Algorithms Made Easy". In: *Journal of Machine Learning Research* 13 (2012), pp. 2171–2175.

[53] T-C. Fu. "A review on time series data mining". In: *Engineering Applications of Artificial Intelligence* 24 (2011), pp. 164–181.

[54] Ahmed Fawzy Gad. "PyGAD: an intuitive genetic algorithm Python library". In: *Multimedia Tools and Applications* 83.20 (2023), pp. 58029–58042. ISSN: 1573-7721. DOI: 10.1007/s11042-023-17167-y. URL: http://dx.doi.org/10.1007/s11042-023-17167-y.

[55] Christopher J. Gaudet and Anthony S. Maida. "Deep quaternion networks". In: *Proceedings of the IEEE International Joint Conference on Neural Networks (IJCNN)*. 2018.

[56] Aurélien Géron. *Hands-on machine learning with Scikit-Learn, Keras, and TensorFlow.* " O'Reilly Media, Inc.", 2022.

[57] David E. Goldberg and John H. Holland. In: *Machine Learning* 3.2/3 (1988), pp. 95–99. ISSN: 0885-6125. DOI: 10.1023/a:1022602019183. URL: http://dx.doi.org/10.1023/A:1022602019183.

[58] Peter J.B. Hancock. "Pruning Neural Nets by Genetic Algorithm". In: *Artificial Neural Networks.* Elsevier, 1992, pp. 991–994. ISBN: 9780444894885. DOI: 10.1016/b978-0-444-89488-5.50036-1. URL: http://dx.doi.org/10.1016/B978-0-444-89488-5.50036-1.

[59] Oğuzhan Hasançebi and Fuat Erbatur. "Evaluation of crossover techniques in genetic algorithm based optimum structural design". In: *Computers & Structures* 78.1–3 (2000), pp. 435–448. ISSN: 0045-7949. DOI: 10.1016/s0045-7949(00)00089-4. URL: http://dx.doi.org/10.1016/S0045-7949(00)00089-4.

[60] Fatma A. Hashim et al. "Archimedes optimization algorithm: a new metaheuristic algorithm for solving optimization problems". In: *Applied Intelligence* 51.3 (2020), pp. 1531–1551. ISSN: 1573-7497. DOI: 10.1007/s10489-020-01893-z. URL: http://dx.doi.org/10.1007/s10489-020-01893-z.

[61] S.S. Haykin. *Neural Networks: A Comprehensive Foundation.* International edition. Prentice Hall, 1999. ISBN: 9780132733502. URL: https://books.google.pl/books?id=bX4pAQAAMAAJ.

[62] M. Holčapek. "Graded equipollence and fuzzy câmeasures of finite fuzzy sets". In: *Proceedings of the IEEE International Conference on Fuzzy Systems (FUZZ-IEEE).* IEEE, 2011, pp. 2375–2382.

[63] John H. Holland. *Adaptation in Natural and Artificial Systems: An Introductory Analysis with Applications to Biology, Control, and Artificial Intelligence.* The MIT Press, 1992. ISBN: 9780262275552. DOI: 10.7551/mitpress/1090.001.0001. URL: http://dx.doi.org/10.7551/mitpress/1090.001.0001.

[64] M. Holčapek et al. "Necessary and sufficient conditions for generalized uniform fuzzy partitions". In: *Fuzzy Sets and Systems* 277 (2015), pp. 97–121.

[65] S. Hongo et al. "Constructing Convolutional Neural Networks Based on Quaternion". In: *2020 International Joint Conference on Neural Networks (IJCNN)*. Glasgow, UK, 2020, pp. 1–6. DOI: 10.1109/IJCNN48605.2020.9207325.

[66] J J Hopfield. "Neural networks and physical systems with emergent collective computational abilities." In: *Proceedings of the National Academy of Sciences* 79.8 (1982), pp. 2554–2558. DOI: 10.1073/pnas.79.8.2554. eprint: https://www.pnas.org/doi/pdf/10.1073/pnas.79.8.2554. URL: https://www.pnas.org/doi/abs/10.1073/pnas.79.8.2554.

[67] K. Hornik, M. Stinchcombe, and H. White. "Multilayer feedforward networks are universal approximators". In: *Neural Networks* 2 (1989), pp. 359–366.

[68] Ali Hosseinalipour et al. "A metaheuristic approach based on coronavirus herd immunity optimiser for breast cancer diagnosis". In: *Cluster Computing* (2024). ISSN: 1573-7543. DOI: 10.1007/s10586-024-04360-3. URL: http://dx.doi.org/10.1007/s10586-024-04360-3.

[69] Cheng-Lung Huang and Chieh-Jen Wang. "A GA-based feature selection and parameters optimizationfor support vector machines". In: *Expert Systems with Applications* 31.2 (2006), pp. 231–240. ISSN: 0957-4174. DOI: 10.1016/j.eswa.2005.09.024. URL: http://dx.doi.org/10.1016/j.eswa.2005.09.024.

[70] G.B. Huang, Q.Y. Zhu, and C.K. Siew. "Multilayer feedforward networks are universal approximators". In: *Neurocomputing* 70 (2006), pp. 489–501.

[71] *Hypercomplex Keras library*. https://pypi.org/project/HypercomplexKeras/. accessed 27 September 2024.

[72] Fernando Itano, Miguel Angelo de Abreu de Sousa, and Emilio Del-Moral-Hernandez. "Extending MLP ANN hyper-parameters Optimization by using Genetic Algorithm". In: *2018 International Joint Conference on Neural Networks (IJCNN)*. IEEE, 2018. DOI: 10.1109/ijcnn.2018.8489520. URL: http://dx.doi.org/10.1109/IJCNN.2018.8489520.

[73] H Joel Jeffrey. "Chaos game representation of gene structure". In: *Nucleic acids research* 18.8 (1990), pp. 2163–2170.

[74] H Joel Jeffrey. "Chaos game visualization of sequences". In: *Computers & Graphics* 16.1 (1992), pp. 25–33.

[75] J. Kacprzyk, A. Wilbik, and S. Zadrożny. "An approach to the linguistic summarization of time series using a fuzzy quantifier driven aggregation". In: *International Journal of Intelligent Systems* 25 (2010), pp. 411–439.

[76] Dervis Karaboga and Bahriye Basturk. "Artificial Bee Colony (ABC) Optimization Algorithm for Solving Constrained Optimization Problems". In: *Foundations of Fuzzy Logic and Soft Computing*. Springer Berlin Heidelberg, pp. 789–798. ISBN: 9783540729501. DOI: 10.1007/978-3-540-72950-1_77. URL: http://dx.doi.org/10.1007/978-3-540-72950-1_77.

[77] J. Kennedy and R. Eberhart. "Particle swarm optimization". In: *Proceedings of ICNN'95 - International Conference on Neural Networks*. ICNN-95. IEEE. DOI: 10.1109/icnn.1995.488968. URL: http://dx.doi.org/10.1109/ICNN.1995.488968.

[78] Serhat KILICARSLAN, Mete CELIK, and Şafak SAHIN. "Hybrid models based on genetic algorithm and deep learning algorithms for nutritional Anemia disease classification". In: *Biomedical Signal Processing and Control* 63 (2021), p. 102231. ISSN: 1746-8094. DOI: 10.1016/j.bspc.2020.102231. URL: http://dx.doi.org/10.1016/j.bspc.2020.102231.

[79] S. Kirkpatrick, C. D. Gelatt, and M. P. Vecchi. "Optimization by Simulated Annealing". In: *Science* 220.4598 (1983), pp. 671–680. ISSN: 1095-9203. DOI: 10.1126/science.220.4598.671. URL: http://dx.doi.org/10.1126/science.220.4598.671.

[80] George J. Klir and Bo Yuan. *Fuzzy Sets and Fuzzy Logic: Theory and Applications*. 574 pp. Upper Saddle River, NJ: Prentice Hall PTR, 1995.

[81] MJ Kochenderfer. *Algorithms for Optimization*. The MIT Press Cambridge, 2019.

[82] A. N. Kolmogorov. "On the representation of continuous functions of many variables by superposition of continuous functions of one variable and addition (in Russian)". In: *Dokl. Akad. Nauk SSSR* 114 (1957), pp. 953–956.

[83] JohnR. Koza. "Genetic programming as a means for programming computers by natural selection". In: *Statistics and Computing* 4.2 (1994). ISSN: 1573-1375. DOI: 10.1007/bf00175355. URL: http://dx.doi.org/10.1007/BF00175355.

[84] Alex Krizhevsky, Ilya Sutskever, and Geoffrey E. Hinton. "ImageNet Classification with Deep Convolutional Neural Networks". In: *Advances in Neural Information Processing Systems*. 2012, pp. 1097–1105.

[85] Puneet Kumar, Shalini Batra, and Balasubramanian Raman. "Deep neural network hyper-parameter tuning through twofold genetic approach". In: *Soft Computing* 25.13 (2021), pp. 8747–8771. ISSN: 1433-7479. DOI: 10.1007/s00500-021-05770-w. URL: http://dx.doi.org/10.1007/s00500-021-05770-w.

[86] S. Kumar and U. Rastogi. "A Comprehensive Review on the Advancement of High-Dimensional Neural Networks in Quaternionic Domain with Relevant Applications". In: *Archives of Computational Methods in Engineering* 30 (2023), pp. 3941–3968. DOI: 10.1007/s11831-023-09925-w. URL: https://doi.org/10.1007/s11831-023-09925-w.

[87] Radosław Kycia and Agnieszka Niemczynowicz. "Hypercomplex neural network in time series forecasting of stock data". In: *arXiv preprint arXiv:2401.04632* (2024).

[88] Yann LeCun et al. "Gradient-Based Learning Applied to Document Recognition". In: *Proceedings of the IEEE* 86.11 (1998), pp. 2278–2324.

[89] Johannes Lederer. "Activation Functions in Artificial Neural Networks: A Systematic Overview". In: *arXiv:2101.09957v1* (2021). URL: https://doi.org/10.48550/arXiv.2101.09957.

[90] Patrick Lewis et al. "Retrieval-augmented generation for knowledge-intensive nlp tasks". In: *Advances in Neural Information Processing Systems* 33 (2020), pp. 9459–9474.

[91] B. Lim and Zohren S. "Time-series forecasting with deep learning: a survey". In: *Phil. Trans. R. Soc. A.* (2021). DOI: http://doi.org/10.1098/rsta.2020.0209.

[92] Z. Liu et al. "KAN: Kolmogorov-Arnold Networks". In: *arXiv: 2404.19756* (2024).

[93]   Hannah F Löchel et al. "Deep learning on chaos game representation for proteins". In: *Bioinformatics* 36.1 (2020), pp. 272–279.

[94]   Hannah Franziska Löchel and Dominik Heider. "Chaos game representation and its applications in bioinformatics". In: *Computational and Structural Biotechnology Journal* 19 (2021), pp. 6263–6271.

[95]   A. Lorke et al. "Cybenko's Theorem and the capability of a neural network as function approximator". In: (2019). URL: https : / / www . mathematik . uni - wuerzburg . de / fileadmin / 10040900 / 2019 / Seminar - Artificial - Neural-Network-24-9-.pdf.

[96]   Junyang Lu et al. "A Survey of Multimodal Large Language Models". In: *arXiv preprint arXiv:2211.12897* (2022).

[97]   Warren S. McCulloch and Walter Pitts. "A logical calculus of the ideas immanent in nervous activity". In: *Bulletin of Mathematical Biophysics* 5.1 (1943), pp. 115–133. DOI: 10 . 1007 / BF02478259. URL: https : / / doi . org / 10 . 1007 / BF02478259.

[98]   Usama Mehboob et al. "Genetic algorithms in wireless networking: techniques, applications, and issues". In: *Soft Computing* 20.6 (2016), pp. 2467–2501. ISSN: 1433-7479. DOI: 10 . 1007/s00500-016-2070-9. URL: http://dx.doi.org/ 10.1007/s00500-016-2070-9.

[99]   Adil Mehdary et al. "Hyperparameter Optimization with Genetic Algorithms and XGBoost: A Step Forward in Smart Grid Fraud Detection". In: *Sensors* 24.4 (2024), p. 1230. ISSN: 1424-8220. DOI: 10 . 3390 / s24041230. URL: http://dx.doi.org/10.3390/s24041230.

[100] Teresa Mendonça et al. "PH^2—A dermoscopic image database for research and benchmarking". In: *2013 35th Annual International Conference of the IEEE Engineering in Medicine and Biology Society (EMBC)*. 2013, pp. 5437–5440.

[101] Seyedali Mirjalili. "The Ant Lion Optimizer". In: *Advances in Engineering Software* 83 (2015), pp. 80–98. ISSN: 0965-9978. DOI: 10 . 1016 / j . advengsoft . 2015 . 01 . 010. URL: http://dx.doi.org/10.1016/j.advengsoft.2015.01. 010.

[102] Bernardo Morales-Castañeda et al. "A better balance in metaheuristic algorithms: Does it exist?" In: *Swarm and Evolutionary Computation* 54 (2020), p. 100671. ISSN: 2210-6502. DOI: 10.1016/j.swevo.2020.100671. URL: http://dx.doi.org/10.1016/j.swevo.2020.100671.

[103] S. A. Morris. "Hilbert 13: Are There any Genuine Continuous Multivariate Real-Valued Functions?" In: *Bulletin of the American Mathematical Society (New Series)* 58/1 (2021), pp. 107–118.

[104] S. A. Morris. "Topology Without Tears". In: (2020). URL: https://www.topologywithouttears.net/topbook.pdf.

[105] G. Moyse and M.J. Lesot. "Linguistic summaries of locally periodic time series". In: *Fuzzy Sets and Systems* 285 (2016), pp. 94–117.

[106] Mohammad Hassan Tayarani Najaran. "An evolutionary ensemble convolutional neural network for fault diagnosis problem". In: *Expert Systems with Applications* 233 (2023), p. 120678. ISSN: 0957-4174. DOI: 10.1016/j.eswa.2023.120678. URL: http://dx.doi.org/10.1016/j.eswa.2023.120678.

[107] *NCBI.* https://www.ncbi.nlm.nih.gov/. accessed 27 September 2024.

[108] H. T. Nguyen and E. A. Walker. *A First Course in Fuzzy Logic.* Boca Raton, Florida: Chapman & Hall/CRC, 2006.

[109] L. Nguyen and V. Novák. "Forecasting Seasonal Time Series Based on Fuzzy techniques". In: *Fuzzy Sets and Systems* 361 (2019), pp. 114–129.

[110] Agnieszka Niemczynowicz and Radosław Antoni Kycia. "Fully tensorial approach to hypercomplex neural networks". In: *arXiv preprint arXiv:2407.00449* (2024).

[111] Agnieszka Niemczynowicz and Radosław Antoni Kycia. "KHNNs: hypercomplex neural networks computations via Keras using TensorFlow and PyTorch". In: *arXiv preprint arXiv:2407.00452* (2024).

[112] V. Novák. "A Comprehensive Theory of Trichotomous Evaluative Linguistic Expressions". In: *Fuzzy Sets and Systems* 159.22 (2008), pp. 2939–2969.

[113] V. Novák. "A Formal Theory of Intermediate Quantifiers". In: *Fuzzy Sets and Systems* 159.10 (2008), pp. 1229–1246.

[114] V. Novák. "Detection of Structural Breaks in Time Series using Fuzzy Techniques". In: *Int. J. of Fuzzy Logic and Intelligent Systems* 18.1 (2018), pp. 1–12.

[115] V. Novák. "Linguistic Characterization of Time Series". In: *Fuzzy Sets and Systems* 285 (2016), pp. 52–72.

[116] V. Novák. "Mining information from time series in the form of sentences of natural language". In: *International Journal of Approximate Reasoning* 78 (2016), pp. 192–209.

[117] V. Novák. "On Fuzzy Type Theory". In: *Fuzzy Sets and Systems* 149 (2005), pp. 235–273.

[118] V. Novák and S. Mirshahi. "On the similarity and dependence of time series". In: *MDPI Mathematics* 9.5 (2021), pp. 550–563. DOI: 0.3390/math9050550. URL: http://www.mdpi.com/2227-7390/9/5/550.

[119] V. Novák, I. Perfilieva, and A. Dvořák. *Insight into Fuzzy Modeling*. Hoboken, New Jersey: Wiley & Sons, 2016.

[120] V. Novák et al. "Analysis of Seasonal Time Series Using Fuzzy Approach". In: *Int. Journal of General Systems* 39 (2010), pp. 305–328.

[121] V. Novák et al. "Filtering out high frequencies in time series using F-transform". In: *Information Sciences* 274 (2014), pp. 192–209.

[122] Hidehiko Okada. "Evolutionary Training of Binary Neural Networks by Differential Evolution". In: *International Journal of Scientific Research in Computer Science and Engineering* 10 (1 Feb. 2022), pp. 26–31. ISSN: 2347-2693. URL: https://www.isroset.org/journal/IJSRCSE/full_paper_view.php?paper_id=2693.

[123] Hidehiko Okada. "Evolutionary Training of Binary Neural Networks by Evolution Strategy". In: *International Journal of Scientific Research in Computer Science and Engineering* 9 (1 Feb. 2021), pp. 32–36. ISSN: 2347-2693. URL: https://www.isroset.org/journal/IJSRCSE/full_paper_view.php?paper_id=2271.

[124] Hidehiko Okada. "Evolutionary Training of Binary Neural Networks by Genetic Algorithm". In: *International Journal of Scientific Research in Computer Science and Engineering* 9 (6 Dec. 2021), pp. 63–68. ISSN: 2347-2693. URL: https://www.isroset.org/journal/IJSRCSE/full_paper_view.php?paper_id=2606.

[125] Sunday O. Oladejo, Stephen O. Ekwe, and Seyedali Mirjalili. "The Hiking Optimization Algorithm: A novel human-based metaheuristic approach". In: *Knowledge-Based Systems* 296 (2024), p. 111880. ISSN: 0950-7051. DOI: 10.1016/j.knosys.2024.111880. URL: http://dx.doi.org/10.1016/j.knosys.2024.111880.

[126] Fatma A. Omara and Mona M. Arafa. "Genetic algorithms for task scheduling problem". In: *Journal of Parallel and Distributed Computing* 70.1 (2010), pp. 13–22. ISSN: 0743–7315. DOI: 10.1016/j.jpdc.2009.09.009. URL: http://dx.doi.org/10.1016/j.jpdc.2009.09.009.

[127] Richard Sheldon Palais. "The classification of real division algebras". In: *The American Mathematical Monthly* 75.4 (1968), pp. 366–368.

[128] Filip Pałka et al. "Hyperspectral Classification of Blood-Like Substances Using Machine Learning Methods Combined with Genetic Algorithms in Transductive and Inductive Scenarios". In: *Sensors* 21.7 (2021), p. 2293. ISSN: 1424-8220. DOI: 10.3390/s21072293. URL: http://dx.doi.org/10.3390/s21072293.

[129] Alejandro Parada-Mayorga and Alejandro Ribeiro. "Algebraic neural networks: Stability to deformations". In: *IEEE Transactions on Signal Processing* 69 (2021), pp. 3351–3366.

[130] Thomas Parcollet et al. "Quaternion convolutional neural networks for end-to-end automatic speech recognition". In: *Proceedings of Interspeech*. 2018.

[131] Adam Paszke et al. "PyTorch: an imperative style, high-performance deep learning library". In: *Proceedings of the 33rd International Conference on Neural Information Processing Systems*. Red Hook, NY, USA: Curran Associates Inc., 2019.

[132] G. Patane. "Fuzzy transform and least-squares approximation: Analogies, differences, and generalizations". In: *Fuzzy Sets and Systems* 180 (2011), pp. 41–54.

[133] Fabian Pedregosa et al. "Scikit-learn: Machine Learning in Python". In: *Journal of Machine Learning Research* 12.85 (2011), pp. 2825–2830. URL: http://jmlr.org/papers/v12/pedregosa11a.html.

[134] I. Perfilieva. "F-transform". In: *Handbook of Computational Intelligence*. Ed. by J. Kacprzyk and W. Pedrycz. San Diego: Springer, 2015, pp. 113–130.

[135] I. Perfilieva. "Fuzzy Transforms: A Challenge to Conventional Transforms". In: *Advances in Images and Electron Physics, 147*. Ed. by P.W. Hawkes. San Diego: Elsevier Academic Press, 2007, pp. 137–196.

[136] I. Perfilieva. "Fuzzy Transforms: theory and applications". In: *Fuzzy Sets and Systems* 157 (2006), pp. 993–1023.

[137] I. Perfilieva, M. Holčapek, and V. Kreinovich. "A new reconstruction from the F-transform components". In: *Fuzzy Sets and Systems* 288 (2016), pp. 3–25.

[138] M. Poluektov and A. Polar. "A new iterative method for construction of the Kolmogorov-Arnold representation". In: *arXiv: 2305.08194* (2023).

[139] Eslam Pourbasheer et al. "Application of genetic algorithm-support vector machine (GA-SVM) for prediction of BK-channels activity". In: *European Journal of Medicinal Chemistry* 44.12 (2009), pp. 5023–5028. ISSN: 0223-5234. DOI: 10.1016/j.ejmech.2009.09.006. URL: http://dx.doi.org/10.1016/j.ejmech.2009.09.006.

[140] Javier Poyatos et al. "EvoPruneDeepTL: An evolutionary pruning model for transfer learning based deep neural networks". In: *Neural Networks* 158 (2023), pp. 59–82. ISSN: 0893-6080. DOI: 10.1016/j.neunet.2022.10.011. URL: http://dx.doi.org/10.1016/j.neunet.2022.10.011.

[141] Alec Radford et al. "Improving Language Understanding by Generative Pre-Training". In: (2018).

[142] Kanchan Rajwar, Kusum Deep, and Swagatam Das. "An exhaustive review of the metaheuristic algorithms for search and optimization: taxonomy, applications, and open challenges". In: *Artificial Intelligence Review* 56.11 (2023), pp. 13187–13257. ISSN: 1573-7462. DOI: 10.1007/s10462-023-10470-y. URL: http://dx.doi.org/10.1007/s10462-023-10470-y.

[143] Fatemeh Ramezani and Shahriar Lotfi. "Social-Based Algorithm (SBA)". In: *Applied Soft Computing* 13.5 (2013), pp. 2837–2856. ISSN: 1568-4946. DOI: 10.1016/j.asoc.2012.05.018. URL: http://dx.doi.org/10.1016/j.asoc.2012.05.018.

[144] Sebastian Raschka, Yuxi Hayden Liu, and Vahid Mirjalili. *Machine Learning with PyTorch and Scikit-Learn: Develop machine learning and deep learning models with Python*. Packt Publishing Ltd, 2022.

[145] Sebastian Raschka and Vahid Mirjalili. *Python machine learning: Machine learning and deep learning with Python, scikit-learn, and TensorFlow 2*. Packt publishing ltd, 2019.

[146] Esmat Rashedi, Hossein Nezamabadi-pour, and Saeid Saryazdi. "GSA: A Gravitational Search Algorithm". In: *Information Sciences* 179.13 (2009), pp. 2232–2248. ISSN: 0020-0255. DOI: 10.1016/j.ins.2009.03.004. URL: http://dx.doi.org/10.1016/j.ins.2009.03.004.

[147] David E Rumelhart, Geoffrey E Hinton, and Ronald J Williams. "Learning representations by back-propagating errors". In: *nature* 323.6088 (1986), pp. 533–536.

[148] David E. Rumelhart, Geoffrey E. Hinton, and Ronald J. Williams. "Learning representations by back-propagating errors". In: *Nature* 323.6088 (1986), pp. 533–536. ISSN: 1476-4687. DOI: 10.1038/323533a0. URL: http://dx.doi.org/10.1038/323533a0.

[149] Dipendra C Sengupta et al. "Similarity studies of corona viruses through chaos game representation". In: *Computational molecular bioscience* 10.3 (2020), p. 61.

[150] Ali Sharifi and Kamal Alizadeh. "Comparison of the Particle Swarm Optimization with the Genetic Algorithms as a Training for Multilayer Perceptron Technique to Diagnose Thyroid Functional Disease". In: *Shiraz E-Medical Journal* 22.1 (2020). ISSN: 1735-1391. DOI: 10.5812/semj.100351. URL: http://dx.doi.org/10.5812/semj.100351.

[151] M. Sharma, J. S. Brownstein, and N. Ramakrishnan. "T3: Domain-Agnostic Neural Time-series Narration". In: *2021 IEEE International Conference on Data Mining (ICDM)*. 2021, pp. 1324–1329. DOI: 10.1109/ICDM51629.2021.00165.

[152] David Sherrington and Scott Kirkpatrick. "Solvable Model of a Spin-Glass". In: *Phys. Rev. Lett.* 35 (26 1975), pp. 1792–1796. DOI: 10.1103/PhysRevLett.35.1792. URL: https://link.aps.org/doi/10.1103/PhysRevLett.35.1792.

[153] Yuhui Shi. "Brain Storm Optimization Algorithm". In: *Advances in Swarm Intelligence*. Springer Berlin Heidelberg, 2011, pp. 303–309. ISBN: 9783642215155. DOI: 10.1007/978-3-642-21515-5_36. URL: http://dx.doi.org/10.1007/978-3-642-21515-5_36.

[154] Anupriya Shukla, Hari Mohan Pandey, and Deepti Mehrotra. "Comparative review of selection techniques in genetic algorithm". In: *2015 International Conference on Futuristic Trends on Computational Analysis and Knowledge Management (ABLAZE)*. IEEE, 2015. DOI: 10.1109/ablaze.2015.7154916. URL: http://dx.doi.org/10.1109/ABLAZE.2015.7154916.

[155] Aman Singh, Nikhil Duggal, and Vibhav Balasubramanian. "Multimodal Large Language Models: A Survey". In: *arXiv preprint arXiv:2304.01868* (2023).

[156] L. Stefanini. "F-transform with parametric generalized fuzzy partitions". In: *Fuzzy Sets and Systems* 180 (2011), pp. 98–120.

[157] Rainer Storn and Kenneth Price. In: *Journal of Global Optimization* 11.4 (1997), pp. 341–359. ISSN: 0925-5001. DOI: 10.1023/a:1008202821328. URL: http://dx.doi.org/10.1023/A:1008202821328.

[158] T. Takagi and M. Sugeno. "Fuzzy identification of systems and its application to modeling and control". In: *IEEE Trans. Syst. Man, Cybern.* 15 (1985), pp. 116–132.

[159] M. H. Tayarani-N and M. R. Akbarzadeh-T. "Magnetic Optimization Algorithms a new synthesis". In: *2008 IEEE Congress on Evolutionary Computation (IEEE World Congress on Computational Intelligence)*. IEEE, 2008. DOI: 10.1109/cec.2008.4631155. URL: http://dx.doi.org/10.1109/CEC.2008.4631155.

[160] Laurence Taylor and Thomas Parcollet. "Applications of quaternion and hypercomplex neural networks in computer vision". In: *Journal of Computational Vision* (2020).

[161] Hugo Touvron et al. "LLaMA: Open and Efficient Foundation Language Models". In: (2023). arXiv: 2302.13971.

[162] P.T.T. Truong and V. Novák. "An Improved Forecasting and Detection of Structural Breaks in Time Series using Fuzzy Techniques". In: *Theory and Applications of Time Series Analysis and Forecasting: Selected Contributions from ITISE 2021*. Ed. by O. Valenzuela et al. Springer, 2023, pp. 3–17.

[163] Chih-Fong Tsai, William Eberle, and Chi-Yuan Chu. "Genetic algorithms in feature and instance selection". In: *Knowledge-Based Systems* 39 (2013), pp. 240–247. ISSN: 0950-7051. DOI: 10.1016/j.knosys.2012.11.005. URL: http://dx.doi.org/10.1016/j.knosys.2012.11.005.

[164] Matteo Valle. *Hypercomplex-valued Convolutional Neural Networks*. GitHub repository. 2020. URL: https://github.com/mevalle/Hypercomplex-valued-Convolutional-Neural-Networks.

[165] Ashish Vaswani et al. "Attention is all you need". In: *Advancs in neural information processing systems* (2017), pp. 5998–6008.

[166] E. Vetrimani, M. Arulselvi, and G. Ramesh. "Building convolutional neural network parameters using genetic algorithm for the croup cough classification problem". In: *Measurement: Sensors* 27 (2023), p. 100717. ISSN: 2665-9174. DOI: 10.1016/j.measen.2023.100717. URL: http://dx.doi.org/10.1016/j.measen.2023.100717.

[167] G. Vieira, M. E. Valle, and W. Lopes. "Clifford Convolutional Neural Networks for Lymphoblast Image Classification". In: *Advanced Computational Applications of Geometric Algebra. ICACGA 2022*. Ed. by D. W. Silva, E. Hitzer, and D. Hildenbrand. Vol. 13771. Lecture Notes in Computer Science. Springer, Cham, 2024. DOI: 10.1007/978-3-031-34031-4_7.

[168] Zhenyu Wang et al. "Network pruning using sparse learning and genetic algorithm". In: *Neurocomputing* 404 (2020), pp. 247–256. ISSN: 0925-2312. DOI: 10.1016/j.neucom.2020.03.082. URL: http://dx.doi.org/10.1016/j.neucom.2020.03.082.

[169] Long Wen et al. "A new genetic algorithm based evolutionary neural architecture search for image classification". In: *Swarm and Evolutionary Computation* 75 (2022), p. 101191. ISSN: 2210-6502. DOI: 10.1016/j.swevo.2022.101191. URL: http://dx.doi.org/10.1016/j.swevo.2022.101191.

[170] A. Wilbik and R. M. Dijkman. "On the generation of useful linguistic summaries of sequences". In: *2016 IEEEInternationla Conference on Fuzzy Systems* (2016), pp. 555–562.

[171] Axel Wismüller et al. "Cluster Analysis of Biomedical Image Time-Series". In: *International Journal of Computer Vision* 46 (2002), pp. 103–128. DOI: 10.1023/A:1013550313321. URL: https://doi.org/10.1023/A:1013550313321.

[172] Thomas Wolf et al. *Transformers: State-of-the-Art Natural Language Processing*. 2020. arXiv: 1910.03771 [cs.CL].

[173] Lingfei Wu et al. "Graph neural networks: foundation, frontiers and applications". In: *Proceedings of the 28th ACM SIGKDD Conference on Knowledge Discovery and Data Mining*. 2022, pp. 4840–4841.

[174] Xin-She Yang. "Firefly Algorithms for Multimodal Optimization". In: *Lecture Notes in Computer Science*. Springer Berlin Heidelberg, 2009, pp. 169–178. ISBN: 9783642049446. DOI: 10.1007/978-3-642-04944-6_14. URL: http://dx.doi.org/10.1007/978-3-642-04944-6_14.

[175] XinâShe Yang and Amir Hossein Gandomi. "Bat algorithm: a novel approach for global engineering optimization". In: *Engineering Computations* 29.5 (2012), pp. 464–483. ISSN: 0264-4401. DOI: 10.1108/02644401211235834. URL: http://dx.doi.org/10.1108/02644401211235834.

[176] Ji-Hyun Yoo et al. "Optimization of Hyper-parameter for CNN Model using Genetic Algorithm". In: *2019 1st International Conference on Electrical, Control and Instrumentation Engineering (ICECIE)*. IEEE, 2019. DOI: 10.1109/icecie47765.2019.8974762. URL: http://dx.doi.org/10.1109/ICECIE47765.2019.8974762.

[177] L. A. Zadeh. "Fuzzy sets". In: *Information and Control* 8 (1965), pp. 338–353.

[178] Dariusz Żelasko, Wojciech Książek, and Paweł Pławiak. "Transmission Quality Classification with Use of Fusion of Neural Network and Genetic Algorithm in Pay& Require Multi-Agent Managed Network". In: *Sensors* 21.12 (2021), p. 4090. ISSN: 1424-8220. DOI: 10.3390/s21124090. URL: http://dx.doi.org/10.3390/s21124090.

[179] Guohui Zhang, Liang Gao, and Yang Shi. "An effective genetic algorithm for the flexible job-shop scheduling problem". In: *Expert Systems with Applications* 38.4 (2011), pp. 3563–3573. ISSN: 0957-4174. DOI: 10.1016/j.eswa.2010.08.145. URL: http://dx.doi.org/10.1016/j.eswa.2010.08.145.

[180] Weiguo Zhao, Liying Wang, and Zhenxing Zhang. "Atom search optimization and its application to solve a hydrogeologic parameter estimation problem". In: *Knowledge-Based Systems* 163 (2019), pp. 283–304. ISSN: 0950-7051. DOI: 10.1016/j.knosys.2018.08.030. URL: http://dx.doi.org/10.1016/j.knosys.2018.08.030.

Index

For Product Safety Concerns and Information please contact our EU
representative GPSR@taylorandfrancis.com
Taylor & Francis Verlag GmbH, Kaufingerstraße 24, 80331 München, Germany

9 781032 847146